Current Topics in Microbiology and Immunology

Volume 360

Current Topics in Microbiology and Immunology

Previously published volumes

Further volumes can be found at www.springer.com

Vol. 329: **Griffin, Diane E.; Oldstone, Michael B. A. (Eds.)**: Measles. 2009.
ISBN 978-3-540-70522-2

Vol. 330: **Griffin, Diane E.; Oldstone, Michael B. A. (Eds.)**:
Measles. 2009.
ISBN 978-3-540-70616-8

Vol. 331: **Villiers, E. M. de (Ed.)**:
TT Viruses. 2009.
ISBN 978-3-540-70917-8

Vol. 332: **Karasev A. (Ed.)**:
Plant produced Microbial Vaccines. 2009.
ISBN 978-3-540-70857-5

Vol. 333: **Compans, Richard W.; Orenstein, Walter A. (Eds.)**:
Vaccines for Pandemic Influenza. 2009.
ISBN 978-3-540-92164-6

Vol. 334: **McGavern, Dorian; Dustin, Micheal (Eds.)**:
Visualizing Immunity. 2009.
ISBN 978-3-540-93862-0

Vol. 335: **Levine, Beth; Yoshimori, Tamotsu; Deretic, Vojo (Eds.)**:
Autophagy in Infection and Immunity. 2009.
ISBN 978-3-642-00301-1

Vol. 336: **Kielian, Tammy (Ed.)**:
Toll-like Receptors: Roles in Infection and Neuropathology. 2009.
ISBN 978-3-642-00548-0

Vol. 337: **Sasakawa, Chihiro (Ed.)**:
Molecular Mechanisms of Bacterial Infection via the Gut. 2009.
ISBN 978-3-642-01845-9

Vol. 338: **Rothman, Alan L. (Ed.)**:
Dengue Virus. 2009.
ISBN 978-3-642-02214-2

Vol. 339: **Spearman, Paul; Freed, Eric O. (Eds.)**:
HIV Interactions with Host Cell Proteins. 2009.
ISBN 978-3-642-02174-9

Vol. 340: **Saito, Takashi; Batista, Facundo D. (Eds.)**:
Immunological Synapse. 2010.
ISBN 978-3-642-03857-0

Vol. 341: **Bruserud, Øystein (Ed.)**:
The Chemokine System in Clinical and Experimental Hematology. 2010.
ISBN 978-3-642-12638-3

Vol. 342: **Arvin, Ann M. (Ed.)**:
Varicella-zoster Virus. 2010.
ISBN 978-3-642-12727-4

Vol. 343: **Johnson, John E. (Ed.)**:
Cell Entry by Non-Enveloped Viruses. 2010.
ISBN 978-3-642-13331-2

Vol. 344: **Dranoff, Glenn (Ed.)**:
Cancer Immunology and Immunotherapy. 2011.
ISBN 978-3-642-14135-5

Vol. 345: **Simon, M. Celeste (Ed.)**:
Diverse Effects of Hypoxia on Tumor Progression. 2010.
ISBN 978-3-642-13328-2

Vol. 346: **Christian Rommel; Bart Vanhaesebroeck; Peter K. Vogt (Ed.)**:
Phosphoinositide 3-kinase in Health and Disease. 2010.
ISBN 978-3-642-13662-7

Vol. 347: **Christian Rommel; Bart Vanhaesebroeck; Peter K. Vogt (Ed.)**:
Phosphoinositide 3-kinase in Health and Disease. 2010.
ISBN 978-3-642-14815-6

Vol. 348: **Lyubomir Vassilev; David Fry (Eds.)**:
Small-Molecule Inhibitors of Protein-Protein Interactions. 2011.
ISBN 978-3-642-17082-9

Vol. 349: **Michael Kartin (Eds.)**:
NF-kB in Health and Disease. 2011.
ISBN 978-3-642-16017-2

Vol. 350: **Rafi Ahmed; Tasuku Honjo (Eds.)**:
Negative Co-receptors and Ligands. 2011.
ISBN 978-3-642-19545-7

Vol. 351: **Marcel, B. M. Teunissen (Ed.)**:
Intradermal Immunization. 2011.
ISBN 978-3-642-23689-1

Vol. 352: **Rudolf Valenta; Robert L. Coffman (Eds.)**
Vaccines against Allergies. 2011.
ISBN 978-3-642-20054-0

Vol. 353: **Charles E. Samuel (Ed.)**:
Adenosine Deaminases Acting on RNA (ADARs) and A-to-I Editing. 2011.
ISBN 978-3-642-22800-1

Vol. 354: **Pamela A. Kozlowski (Ed.)**:
Mucosal Vaccines. 2012.
ISBN 978-3-642-23692-1

Vol. 355: **Ingo K. Mellinghoff; Charles L. Sawyers (Eds.)**:
Therapeutic Kinase Inhibitors, 2012.
ISBN 978-3-642-28295-9

Vol. 356: **Cornelis Murre (Ed.)**:
Epigenetic Regulation of Lymphocyte Development. 2012.
ISBN 978-3-642-24102-4

Vol. 357: **Nicholas J. Mantis (Ed.)**:
Ricin and Shiga Toxins. 2012.
ISBN 978-3-642-27469-5

Vol. 359: **Benhur Lee; Paul Rota(Eds.)**:
Henipavirus. 2012.
ISBN 978-3-642-29818-9

Freddy Radtke
Editor

Notch Regulation of the Immune System

Responsible series editor: Tasuku Honjo

Springer

Editor
Freddy Radtke
EPFL SV ISREC UPRAD
Ecole Polytechnique Fédérale de Lausanne
Lausanne
Switzerland

ISSN 0070-217X
ISBN 978-3-642-24293-9 ISBN 978-3-642-24294-6 (eBook)
DOI 10.1007/978-3-642-24294-6
Springer Heidelberg New York Dordrecht London

Library of Congress Control Number: 2012943368

Printed on acid-free paper

Springer is part of Springer Science+Business Media (www.springer.com)

Preface

From the Cradle to the Work Place: Pleitropic Roles of Notch in Immunity

The Notch signaling cascade is one of the few basic cellular mechanisms that controls cellular fate specifications, cell growth and cellular differentiation. The pathway is highly pleiotropic and regulates multiple aspects of embryogenesis as well as differentiation processes and homeostasis in adult tissues. One of the first reports linking the Notch pathways with the hematopoietic system was the landmarking study in 1991 by Ellisen et al. who associated altered Notch function with human disease identifying a rare chromosomal translocation in human T cell acute lymphoblastic leukemia (T-ALL) (Ellisen et al. 1991). Subsequently, more than a decade ago the essential role of Notch1 for T lymphopoiesis was established (Radtke F. et al. 1999, Pui J. C. et al. 1999). Since then, Notch signaling has been implicated in many aspects of hematopoiesis and immune function. This collection of reviews summarizes the different roles of Notch signaling within the blood system. Bigas and colleagues review our current understanding of Notch function for the generation, specification and maintenance of hematopoietic stem cells during embryonic development and adulthood. The best-studied function of Notch signaling within the blood system is its essential role during T cell lineage commitment and maturation. Shah and Zuniga-Pflücker focus on the role of the Notch ligands and key molecules involved in ligand endocytosis as well as on the mechanism of Notch activation during T lymphopoiesis. Toribio and colleagues describe the interplay between Notch and cytokine signaling with a particular emphasis on how Notch-driven expression of the IL-7R is important for survival and proliferation during early thymocyte development. Moreover, they explain how aberrant IL-7R expression or gain of function mutations within the IL-7R gene contribute to T cell neoplasms.

Most of our current knowledge of Notch function within the hematopoietic system is derived from studies in mouse models. However, a major breakthrough was made with the establishment of Notch ligand expressing stromal cell culture systems (Schmitt and Zuniga-Pflucker 2002) that made it possible to now study in vitro also the effects of Notch signaling in human progenitors. These studies revealed the importance of Notch signaling during multiple stages of T cell development, in both mice and men, but they also highlighted species related differences. Taghon and colleagues summarize these findings here for us.

Once thymic T cell development is completed, T cells leave the thymus and orchestrate immunity against pathogens. After receiving and integrating the appropriate signals, T cells differentiate into effector or helper T cell subsets. Different aspects of the role of Notch signaling in T helper cell differentiation and function are described in two reviews by Minter and Osborne and by Tachini-Cottier and colleagues.

More recent evidence also suggests a role for the Notch cascade in alloreactive T cell immunity, which mediates recognition of foreign antigens in recipients of organ transplants. The review by Chung and Maillard highlights these novel findings and they suggest that modulating Notch signaling in T cells could have beneficial effects in an allogeneic transplantation setting.

The family of Notch receptors consists of four family members, of which Notch1, Notch2 and to a lesser extent Notch3 have been shown to mediate physiological relevant functions within the immune system. The review by Sakata-Yanagitoma and Chiba focuses on the specific roles of Notch2 during hematopoietic lineage specification and immune function.

Constitutive Notch1 signaling and activating mutations in the Notch1 gene contribute to and are a major hallmark of T cell lymphoblastic leukemia (T-ALL). The last review by Tzoneva and Ferrando summarizes the recent progress on our understanding of how aberrant Notch1 signaling mediates its oncogenic properties and discusses potential avenues targeting Notch1 therapeutically to fight T-ALL.

Lausanne, Switzerland Freddy Radtke

References

Ellisen LW et al (1991) TAN-1, the human homolog of the Drosophila notch gene, is broken by chromosomal translocations in T lymphoblastic neoplasms. Cell 66(4):649–661

Pui JC et al (1999) Notch1 expression in early lymphopoiesis influences B versus T lineage determination. Immunity 11(3):299–308

Radtke F et al (1999) Deficient T cell fate specification in mice with an induced inactivation of Notch1. Immunity 10(5):547–558

Schmitt TM, Zuniga-Pflucker JC (2002) Induction of T cell development from hematopoietic progenitor cells by delta-like-1 in vitro. Immunity 17(6):749–756

Contents

Contributors

Juan Alcain Centro de Biología Molecular "Severo Ochoa" Consejo Superior de Investigaciones Científicas, Universidad Autónoma de Madrid, 28049 Madrid, Spain

Floriane Auderset Department of Biochemistry, WHO Immunology Research and Training Center, University of Lausanne, Chemin Des Boveresses 155, 1066 Epalinges, Switzerland

Anna Bigas Program in Cancer Research. Institut Mar Investigacions Mèdiques (IMIM), Hospital del Mar, Parc de Recerca Biomèdica de Barcelona, Dr. Aiguader 88, 08003 Barcelona, Spain, e-mail: abigas@imim.es

Shigeru Chiba Department of Hematology, Faculty of Medicine, University of Tsukuba, 1-1-1 Tennodai, Tsukuba, Ibaraki 305-8575, Japan, e-mail: schiba-tky@umin.net

Jooho Chung Life Sciences Institute, University of Michigan, Ann Arbor, MI, USA; Cellular and Molecular Biology Program, University of Michigan, Ann Arbor, MI, USA

Manuel Coutaz Department of Biochemistry, WHO Immunology Research and Training Center, University of Lausanne, Chemin Des Boveresses 155, 1066 Epalinges, Switzerland

Teresa D'Altri Program in Cancer Research. Institut Mar Investigacions Mèdiques (IMIM), Hospital del Mar, Parc de Recerca Biomèdica de Barcelona, Dr. Aiguader 88, 08003 Barcelona, Spain

Lluis Espinosa Program in Cancer Research. Institut Mar Investigacions Mèdiques (IMIM), Hospital del Mar, Parc de Recerca Biomèdica de Barcelona, Dr. Aiguader 88, 08003 Barcelona, Spain

Adolfo A. Ferrando Departments of Pediatrics and Pathology and Institute for Cancer Genetics, Columbia University Medical Center, New York, NY 10032, USA, e-mail: af2196@columbia.edu

Marina García-Peydró Centro de Biología Molecular "Severo Ochoa" Consejo Superior de Investigaciones Científicas, Universidad Autónoma de Madrid, 28049 Madrid, Spain

Sara González-García Centro de Biología Molecular "Severo Ochoa" Consejo Superior de Investigaciones Científicas, Universidad Autónoma de Madrid, 28049 Madrid, Spain

Ivan Maillard Life Sciences Institute, University of Michigan, Ann Arbor, MI, USA; Cellular and Molecular Biology Program, University of Michigan, Ann Arbor, MI, USA; Department of Medicine, Division of Hematology-Oncology, Philadelphia, PA, USA; Department of Cell and Developmental Biology, University of Michigan, Ann Arbor, MI 48104, USA, e-mail: imaillar@umich.edu

Lisa M. Minter Department of Veterinary and Animal Sciences, Program in Molecular and Cellular Biology, University of Massachusetts, Amherst, MA 01003, USA, e-mail: lminter@vasci.umass.edu

Barbara A. Osborne Department of Veterinary and Animal Sciences, Program in Molecular and Cellular Biology, University of Massachusetts, Amherst, MA 01003, USA, e-mail: osborne@vasci.umass.edu

Mamiko Sakata-Yanagimoto Department of Hematology, Faculty of Medicine, University of Tsukuba, 1-1-1 Tennodai, Tsukuba, Ibaraki 305-8575, Japan

Divya K. Shah Department of Immunology, Sunnybrook Research Institute, University of Toronto, 2075 Bayview Avenue, Toronto, ON M4 N 3M5, Canada, e-mail: dkshah@sri.utoronto.ca

Fabienne Tacchini-Cottier Department of Biochemistry, WHO Immunology Research and Training Center, University of Lausanne, Chemin Des Boveresses 155, 1066 Epalinges, Switzerland, e-mail: Fabinne.Tacchini-Cottier@unil.ch

Tom Taghon Department of Clinical Chemistry, Microbiology and Immunology, Ghent University Hospital, Ghent University, 9000 Ghent, Belgium, e-mail: tom.taghon@ugent.be

María L. Toribio Centro de Biología Molecular "Severo Ochoa" Consejo Superior de Investigaciones Científicas, Universidad Autónoma de Madrid, 28049 Madrid, Spain, e-mail: mtoribio@cbm.uam.es

Gannie Tzoneva Institute for Cancer Genetics and Graduate Program in Pathobiology and Molecular Medicine, Columbia University Medical Center, New York, NY 10032, USA

Els Waegemans Department of Clinical Chemistry, Microbiology and Immunology, Ghent University Hospital, Ghent University, 9000 Ghent, Belgium

Inge Van de Walle Department of Clinical Chemistry, Microbiology and Immunology, Ghent University Hospital, Ghent University, 9000 Ghent, Belgium

Juan Carlos Zúñiga-Pflücker Department of Immunology, Sunnybrook Research Institute, University of Toronto, 2075 Bayview Avenue, Toronto, ON M4 N 3M5, Canada, e-mail: jczp@sri.utoronto.ca

The Notch Pathway in Hematopoietic Stem Cells

Anna Bigas, Teresa D'Altri and Lluis Espinosa

Abstract Hematopoiesis is the process that generates all the cell types of the blood, which are responsible for oxygen transport and immune defense. It has been now more than 50 years from the demonstration that blood cells derive from a common ancestor called Hematopoietic Stem Cell (HSC) McCulloch and Till (1960). Thus, the hematopoietic process relies on the unlimited and distinctive self-renewal ability of HSC, which in the adult mammalian organisms reside in the bone marrow, but their generation occurs during embryonic life. Questions still remain about how HSCs acquire and maintain the features of self-renewal and pluripotency that define stem-cell populations. Notch is a crucial signaling pathway involved in the generation of cell diversity and stem-cell maintenance in different systems. In some cases, Notch prevents differentiation, while in other contexts Notch directly participates in promoting cell differentiation. In the following sections, we will review what is known about the role of Notch in HSC establishment and hematopoietic cell lineage specification.

Contents

A. Bigas (✉) · T. D'Altri · L. Espinosa
Program in Cancer Research. Institut Mar Investigacions Mèdiques (IMIM), Hospital del Mar, Parc de Recerca Biomèdica de Barcelona, Dr. Aiguader 88, 08003 Barcelona, Spain
e-mail: abigas@imim.es

Current Topics in Microbiology and Immunology (2012) 360: 1–18
DOI: 10.1007/82_2012_229

Published Online: 14 June 2012

1 The Basics of the Notch Signaling Pathway

Notch is a single transmembrane receptor, which activation depends on cell–cell interactions. Association of the Notch ligand (expressed in the "sending" cell) with the Notch receptor expressed in a neighboring cell ("receiving" cell) induces the proteolytic cleavage of Notch protein, resulting in the release of its intracellular domain (Notch-IC). Then Notch-IC translocates to the nucleus where it associates with the DNA-binding protein RBPj/CSL and the coactivator Mastermind. Formation of this nuclear complex is the bases for the productive transcription of different Notch-targets such as Hairy/Enhancer of Split (Hes) family of genes [reviewed in (Kopan and Ilagan 2009)]. The Hes proteins include several transcriptional repressors involved in most of the Notch functions in cell fate specification (Kageyama et al. 2007). In mammals, there are four Notch receptors (Notch1–4) and five canonical ligands (Jagged1–2 and Delta1, 3, 4), being most of them required during embryonic development, indicative of their unique and specific functions (with the exception of Notch4) [reviewed in (Wang 2011)]. Efficient Notch activation also requires the endosomal internalization of the ligands in the sending cell (before they are presented at the cellular membrane), in a process regulated through ubiquitination by the E3-ubiquitin ligases Mindbomb and Neuralized (Le Borgne et al. 2005). Thus, Mindbomb deficiency results in defective Notch activation, in both invertebrates and vertebrates (Itoh et al. 2003; Koo et al. 2005).

Another important step in the regulation of Notch signaling is the glycosylation of the receptor that occurs in the Golgi apparatus. One of the enzymes that glycosylates Notch is Pofut-1 that catalyzes the O-fucosylation of specific EGF-like repeats in the extracellular domain of the receptor. Pofut-mediated fucosylation is required for the subsequent modification of Notch by Fringe. There are three different Fringe homologs (Lunatic, Manic and Radical Fringe) that are responsible to incorporate *N*-acetylglucosamine to the *O*-fucose group added by Pofut-1. Different Fringe homologs modify specific EGF-like repeats with distinct efficiencies (Wang et al. 2001). Fringe-modifications enhance the capacity of Notch to be activated by ligands of the Delta-like family (3 and 4), whereas reduce Notch activation by the Serrate/Jagged ligands (Jag1–2) (Yang et al. 2005).

2 HSC Generation in the Embryo

It is now well established that HSCs are formed during embryonic life. This process totally depends on Notch activity, since several Notch loss-of-function mutants, including Notch1, RBPj, Mindbomb, and Jagged1 show impaired embryonic hematopoietic development [reviewed in (Bigas et al. 2010)]. The mechanisms regulating HSC generation are not yet elucidated and many different questions are still under investigation. For example, where are HSC formed or which are the embryonic tissues capable of HSCs specification? In fact, Yolk Sac (YS), Aorta/Gonad/Mesonephros (AGM), Placenta, and Fetal liver all contain HSC activity at specific stages of embryonic development, but the identification of the precursors that give rise to HSCs is still under debate.

2.1 *Different Waves of Hematopoietic Differentiation Occur During Embryonic Life*

Several waves of hematopoietic development take place during embryonic life. In the mouse embryo, three waves are clearly defined: primitive, transient definitive, and definitive [reviewed by (Dzierzak and Speck 2008)]. The first hematopoietic cells in the mouse embryo originate around embryonic day 7.5 (E7.5) in the yolk sac. These cells arise from the extraembryonic mesodermal layer concomitantly with the surrounding endothelial layer and concentrate in a continuous belt around the yolk sac, that was previously thought to be composed of multiple island-like structures known as blood islands (Ferkowicz and Yoder 2005). Blood islands contain maturing primitive erythrocytes, characterized by their big size, the presence of nucleus and by the expression of fetal hemoglobin (Barker 1968). These cells represent the predominant hematopoietic population generated during the yolk sac stage, in addition to some macrophages, megakaryocytes, and mast cells (Palis et al. 1999). However, there is no evidence for HSC activity during this primitive hematopoietic wave. In contrast, YS cells from E9 embryos are able to reconstitute busulfan-treated newborn mice, indicating that HSC activity is already present (Fraser et al. 2002). The fact that circulation between the YS and the embryo proper (through the vitelline vessels) is established around E8.5, has hindered the identification of the tissue of origin HSCs. Further increasing the controversy about the origin of these cells during embryonic development, the second transient wave of YS hematopoiesis, following this early primitive wave, generates multipotent erythro-myeloid progenitors (EMPs), neutrophils, and definitive erythrocyte (McGrath et al. 2011). Studies in the mouse embryo show that EMPs and multipotent progenitors express similar cell surface markers (CD45-/low cKit+ CD41+) but mainly differ in their differentiation and clonogenic potential (Bertrand et al. 2005). These YS progenitors colonize the blood stream of the embryo and the fetal liver, and coexist with hematopoietic cells formed in the AGM region and with their progeny.

Although the possibility that YS cells contain some HSC activity has not totally been discarded, increasing evidence from different experimental systems support the current view that adult HSCs are autonomously generated in the embryo proper. Pioneering grafting experiments performed by Dietèrlen-Lievre using quail-chick embryo chimeras (Dieterlen-Lievre et al. 1976) nicely showed that all cell lineages in the adult blood were originated in the embryo proper and not in the extraembryonic yolk sac. Later on, Roger Patient's group using Xenopus as a model, elegantly demonstrated that primitive and definitive hematopoiesis arises from different blastomers after the 32-cell stage (Ciau-Uitz et al. 2000). Moreover, they found that both type of cells are generated from endothelial precursors but they have different origins (Walmsley et al. 2002).

In 1994, Dzierzak and colleagues (Muller et al. 1994) demonstrated that in the E11.5 mouse embryo the Aorta/Gonad/Mesonephros (AGM) is the only tissue capable of reconstituting irradiated mice while little HSC activity was found in the murine YS at this stage. Recently, transplantation experiments using human AGM, YS and placenta (Ivanovs et al. 2011) indicated that the AGM is the main source of HSCs in the human embryo, (Ivanovs et al. 2011), which is in agreement with previous results obtained in the mouse (Gekas et al. 2005; Ottersbach and Dzierzak 2005). However, following this first peak of HSC activity in AGM (around day E11 in the mouse), the placenta becomes the major hematopoietic organ showing HSC activity in the embryo (Gekas et al. 2005), though there is no evidence that the placenta can autonomously generate HSC. Wherever they are formed, after the AGM and placenta stages, HSCs migrate to the fetal liver, where they are amplified before reaching their definitive niche in the bone marrow. Other specialized hematopoietic organs including the thymus and the spleen are also colonized by the corresponding progenitors close to birth, similar to the bone marrow (Cumano and Godin 2007).

2.2 *Origin of HSCs: Hemangioblast, Hemogenic Endothelium, Arterial Endothelium, or Mesodermal Endothelial-like Precursor*

Cell tracing experiments consistently demonstrated that murine adult hematopoietic cells originate from embryonic cells that express Vascular Endothelial Cadherin (VEC) and Runx1 (Chen et al. 2009; Zovein et al. 2008). Since, VEC is a component of the adherent junctions of endothelial cells, these results strongly suggested an endothelial embryonic origin of the HSCs. Moreover, autonomous induction of VEC-CRe-ERT from YS, AGM, or fetal liver explants followed by transplantation experiments demonstrated that only endothelial-like AGM cells are able to reconstitute adult hematopoiesis while endothelial cells from the YS and FL does no longer generate HSCs indicating that the presumptive VEC+ progenitor is restricted both temporally and spatially in the embryo. However, some

VEC can be detected in fetal liver HSCs both in mouse and human likely correspond to cells that have migrated from the AGM but they no longer express the VEC gene (Oberlin et al. 2010). The possibility that HSC are generated from non-endothelial precursors closely associated to the endothelium that expresses VEC is still under study (Bertrand et al. 2005; de Bruijn et al. 2002; Rybtsov et al. 2011).

On the other hand, and supporting the notion that hematopoietic and endothelial cells share a common endothelial-like ancestor called hemangioblast (Eilken et al. 2009; Lancrin et al. 2009), single Embryonic Stem Cells (ESC) have been successfully induced to differentiate into both lineages. However, the physiological relevance of this finding is controversial since ESC cannot generate HSC without previous genetic manipulation (Wang et al. 2005b). To reconcile most of the current knowledge the term "hemogenic endothelium" has been introduced to designate cells with HSC capacity and endothelial appearance. Thus, hemogenic endothelium is the tissue that contains HSC precursors, but whether they originate from the endothelium itself or from a mesodermal cell precursor with or without endothelial capacity is still unknown. In any case, the hemogenic endothelium shows characteristics of arterial endothelium, which suggests that acquisition of the arterial fate is a prerequisite for HSC generation.

2.3 Notch is Dispensable to Generate Primitive or Transient Definitive Hematopoietic Cells

Mouse mutant for Notch1 and RBPj contain normal numbers of primitive erythrocytes and Ery-P progenitors (Kumano et al. 2003; Robert-Moreno et al. 2005), indicating that early YS hematopoiesis is Notch-independent. In fact, RBPj embryos show a higher production of primitive erythrocytes likely as a consequence of the reduced oxygen concentrations that occur in these embryos due to the defects in allantoid fusion (Oka et al. 1995; Robert-Moreno et al. 2007). Similarly, the second wave of YS hematopoiesis that generates all types of myeloid precursors, is also normal in the RBPj- and the Notch1-deficient embryos (Kumano et al. 2003; Robert-Moreno et al. 2007). Further evidence for the Notch-independent formation of transient yolk sac hematopoiesis came from zebrafish studies. Similar to the mammals, zebrafish primitive hematopoiesis is generated in the blood islands and EMP progenitors are characterized by expression of Lmo2. Using reporter models, it was shown that both primitive erythrocytes and transient definitive progenitors are normally formed in Notch-defective Mindbomb mutants (Bertrand et al. 2010; Burns et al. 2005). Finally, using a model of mouse chimeras, it was demonstrated that Notch1 mutant cells contribute to YS and fetal liver hematopoiesis but not to adult definitive hematopoiesis (Hadland et al. 2004).

2.4 Hematopoietic Waves Occurring in Embryonic Stem-Cell Cultures

Under specific culture conditions, ESCs develop into embryoid bodies that generate all types of hematopoietic cells and progenitors (Choi et al. 1998). However, after many years of intense investigation robust hematopoietic reconstitution from non-manipulated ESC-derived cells has not been achieved (Ledran et al. 2008; Wang et al. 2005a), although HSC activity can be obtained by introduction of HoxB4 or cdx4 genes (Wang et al. 2005b). A similar approach, that is ectopic expression of HoxB4, was successfully used to transform yolk sac cells into long-term reconstituting cells (Kyba et al. 2002). Together, these results suggest that ESC hematopoietic differentiation mainly recapitulates the transient embryonic yolk sac hematopoiesis instead of the adult definitive one (Zambidis et al. 2005). On the other hand, it was demonstrated that using specific combinations of growth factors ESC generate Flk1-derived cells resembling AGM/definitive HSCs (coexpressing CD41, sca, c-kit and AA4.1) (Irion et al. 2010), thus indicating the possibility to obtain HSCs in vitro.

During ESC differentiation into hematopoietic commitment, Notch activity is first required to initiate differentiation into all three embryonic layers (mesoderm, ectoderm and endoderm). Then, Hes expression is transiently required during the first stages of EB-differentiation for further hematopoietic differentiation (Yu et al. 2008). During the first wave of ESC-derived hematopoiesis, Notch functions as a negative regulator of primitive erythropoiesis. Notch signaling needs to be silenced to achieve correct primitive erythroid progenitor development (Cheng et al. 2008). Unexpectedly, forced Notch4 activation in the ESC-derived hemangioblasts, defined as Brachyury+Flk1+, induced the formation of cardiac mesoderm (Chen et al. 2008; Schroeder et al. 2003). These results reflect the complexity of Notch signaling during the ESC differentiation process, and suggest that different pulses of specific Notch doses are required for the correct hematopoietic specification of this embryonic cell population mimicking particular stages of embryonic development. Thus, reproducing normal HSC differentiation from ESC will require the fine coordination of Notch with other factors such as Wnt and BMP that are currently being used in ESC differentiation protocols (Nostro et al. 2008).

2.5 Notch is Required for HSC Generation

As mentioned in previous sections of this review, there is growing evidence that definitive HSC are generated in the aortic endothelium of the AGM. Studies from different organisms have clearly demonstrated that HSC emergence is closely associated to the arterial vessels, and mainly but not exclusively to the aorta (de Bruijn et al. 2000). Thus, mutants with altered arterial program fail to produce HSCs, whereas impairment of the venous program leads in some cases to ectopic hematopoietic cell generation. For example, mouse mutants for Activin A-receptor typeII-like1 (acvrl1, alk1) or endoglin (CD150) show ectopic hematopoiesis in veins or vein-like vessels that do not express the arterial marker EphrinB2 (Sorensen et al. 2003; Urness et al. 2000; You et al. 2005). Notch is unequivocally involved in the establishment of the arterial program and different mouse and zebrafish mutants such as RBPj, Notch1, Delta-like4 or Mindbomb cannot initiate this program (Duarte et al. 2004; Koo et al. 2005; Krebs et al. 2004; Lawson et al. 2001). Moreover, induction of Coup-tfii in the venous progenitors is required to inhibit Notch and to allow venous differentiation (You et al. 2005). This fact has extremely hampered the identification of a putative Notch function in HSC formation independently of its role in the arterial specification. Interestingly, experiments conducted in zebrafish showed that ectopic expression of Notch-IC in venous vessels gave rise to hematopoietic Runx1 positive cells in the absence of the arterial marker EfnB2, suggesting that forced Notch activity was sufficient to initiate the hematopoietic program (Burns et al. 2005). However, the functional capacity of these HSCs has not been yet tested. Recently, characterization of Jagged mutant embryos has demonstrated that Notch activation downstream of Jagged1 is specifically required for hematopoietic development without affecting the arterial program (Robert-Moreno et al. 2008) that mainly depends on Delta-like4 signaling (Duarte et al. 2004; Krebs et al. 2004). Notch modifiers such as Mindbomb1 are also important not only for HSC specification but also for regulating the arterial development. Mindbomb was originally thought to be required for Delta-like signaling (Itoh et al. 2003) but recent studies showed its involvement in Jagged signaling (Yamamoto et al. 2010). The principal Notch homolog involved in the hematopoietic development is Notch1, and early studies from Hirai's lab excluded any involvement of Notch2 in HSC emergence. Moreover, ectopic expression of active Notch1 was sufficient to rescue the hematopoietic phenotype imposed by γ-secretase inhibition (Kumano et al. 2003). However, multiple questions remain unknown including the identity of the Notch sending and receiving cells in this process or the developmental stages when Notch is required for HSC formation.

3 Adult Hematopoietic Stem Cells and Progeny

3.1 Self-Renewal of HSC and Asymmetric Cell Division: A Role for Notch?

During the adult life of the mammalian organisms, hematopoietic cells are renewed with a period that varies depending on the cell type (from 5 to 120 days). However, and as we already explained, HSC are generated in the AGM, before their amplification in the fetal liver and colonization of the bone marrow. Thus, the adult bone marrow is not a place of HSC generation but constitutes the niche that

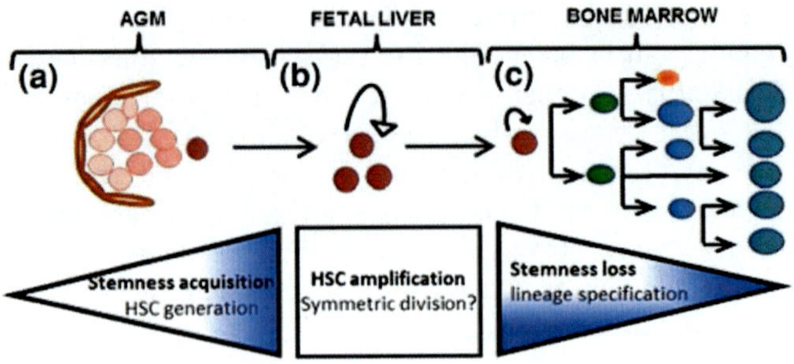

Fig. 1 Model for acquisition and maintenance of stemness in the hematopoietic system. **a** HSCs are originated during embryonic life in the hematopoietic clusters of the AGM. Notch is required during this process, although the target cell and the molecular mechanism are unknown. **b** HSCs migrate to Fetal liver where amplification of this population occurs. There is no evidence for a Notch function during this process. These cells will migrate to bone marrow before birth. **c** Bone marrow is the niche for adult HSCs. HSCs will maintain their identity and generate differentiated cells by asymmetric cell division. Notch dosage controls cell fate acquisition: high Notch is required for T-lymphocyte differentiation, whereas lower levels of Notch are permissive for myeloid differentiation. Increase or decrease of stemness capacity is represented below and blue gradients indicate different requirements for Notch activity

allows these cells to self-renew throughout life and at the same time to differentiate into all different hematopoietic cell types, when necessary (Fig. 1). How the equilibrium between self-renewal and differentiation into the appropriate lineages is achieved remains poorly understood. The most accepted model is that individual HSCs divide asymmetrically to produce a copy of itself and a non-stem cell progenitor already committed to differentiation. Although there is no conclusive evidence of this mechanism, asymmetrical localization of endosomes and key components in the HSCs has been demonstrated (Beckmann et al. 2007; Giebel et al. 2006). Studies from invertebrates indicate that stem-cell polarization is an intrinsic mechanism that involves the asymmetric localization of specific protein complexes and the specific orientation of the mitotic spindle (Furthauer and Gonzalez-Gaitan 2009). Once the cell division has been completed, the two stem-cell daughters can signal each other through specific pathways such as Notch. Signaling through Notch regulates the maintenance of stem-cell populations in several tissues. Moreover, the strength of activation can impinge in the destiny of the stem-cell daughters. In Drosophila, this is the case for intestinal stem cells that differentiate into absorptive or secretory lineages depending on Notch dosage (Ohlstein and Spradling 2007). In addition to asymmetric distribution of Notch receptors and ligands, the inheritance of negative regulators of the pathway such as Numb is also crucial in the maintenance of stem cells in the neural population (Zhong et al. 1996) or in the muscle satellite cells (Conboy and Rando 2002). In the Drosophila sensory organ precursors, Numb is inherited asymmetrically during cell division, leading to enhanced endocytosis-mediated depletion of the surface-available Notch receptor in the cell carrying high Numb and activation of Notch signaling in the other (Couturier et al. 2012). Although, there is no conclusive data indicating that these mechanisms play a role in HSC generation, real-time imaging demonstrated that Numb is asymmetrically distributed after HSC and Lin-Sca+Kit+cd34- cell division in vitro. Consistent with the role of Numb as inhibitor of Notch signaling, cells inheriting high Numb levels showed reduced activation of a Notch-dependent reporter (Wu et al. 2007). Another protein that regulates the Notch pathway is Musashi (Msi). Msi is an RNA-binding protein that interferes with translation, being Numb one of its best-known substrates. Msi2 is required to maintain the reconstitution potential of HSC (Kharas et al. 2010) and specific progenitors (de Andres-Aguayo et al. 2011), however a direct link between Msi2 and Notch activity in these cells has not been established.

3.2 *Notch is not Required for Adult Bone Marrow HSC Maintenance In Vivo*

Different studies aimed to address the role of Notch in the adult HSC compartment. Using a conditional deletion of Notch1 in the murine hematopoietic system (induced by the interferon-regulated Cre recombinase, Mx-cre), it was first demonstrated its essential requirement for early T-cell progenitors to progress throughout T-cell differentiation, whereas other hematopoietic lineages remained unaffected. Most importantly, the reconstitution capacity of HSCs was maintained intact after Notch1 deletion, as shown in competitive transplantation experiments (Radtke et al. 1999). These experiments were a breakthrough in the field, although Notch1 is not the only Notch homolog that can impact in HSC function. Since these early experiments, different hematopoietic-specific Notch mutant mice have been generated and their analysis has confirmed that long-term adult bone marrow HSC functions are Notch-independent. These loss-of-function hematopoietic mutants include Notch2 (Varnum-Finney et al. 2011), Jagged1 (Mancini et al. 2005), RBPj and dominant negative Maml-1 (Maillard et al. 2008), and Mindbomb (Kim et al. 2008). These different models consistently support the idea that Notch activity is no longer needed by adult mammalian HSC after their generation. Surprisingly, it has been recently demonstrated that Notch2-deficient HSC/progenitor cells show a reduced capacity to reconstitute irradiated mice when transplanted in limiting dilution assays as well as to restore non-transplanted animals treated with 5-FU (Varnum-Finney et al. 2011). Together, these data indicate that Notch is not required for HSC maintenance in the mouse, at least under physiological conditions, however it can be crucial under specific stress conditions to maintain the homeostasis of the system. It fact, modulating Notch signal can be also useful for clinical applications, such as ex vivo expansion of HSCs that involves certain doses of cellular stress and forced proliferation.

3.3 Notch Activation Results in Expansion of Hematopoietic Progenitors

Constitutive activation of Notch1 in murine undifferentiated bone marrow cells (Lin-Sca+Kit+) produced a cytokine-dependent immortalized cell line that was able to self-renewal and differentiate into myeloid and lymphoid lineages (Varnum-Finney et al. 2000). From these results, it was further demonstrated that activation of Notch in bone marrow progenitors by immobilized ligands is a powerful tool for expanding HSCs for clinical applications (Stier et al. 2002; Varnum-Finney et al. 2003). In vivo, Jagged ligands are expressed by different components of the bone marrow stroma, including the endothelial and the osteoblastic niche (Calvi et al. 2003). Nowadays, optimized culture conditions allow the efficient expansion of short-term lymphoid and myeloid progenitors from samples, such as cord blood, that contain low numbers of these specific cell populations and transplantation protocols using cord blood as a source of HSC are under clinical investigation in immune-ablated patients. In this clinical trials, individual patients are transplanted with one unmanipulated and one Notch-expanded cord blood unit, and the first results indicate that this protocol reduced the time of neutrophil engraftment compared with patients transplanted with two unmanipulated cord blood units (Delaney et al. 2010). However, identification of the crucial elements that support Notch activation in the HSC in vivo, normally supplied by different components of the bone marrow stroma, will provide crucial information to improve the expansion protocols and obtain better long-term engraftment of the manipulated units. At the mechanistic level, the impact of Notch activation on ex vivo expansion of LT- or ST-HSC may directly correlate with its capacity to regulate cell cycle and self renewal of the HSC population (Carlesso et al. 1999).

3.4 Notch in Hematopoietic Differentiation and Hematologic Disorders

The human Notch1 ortholog was first cloned from a T-cell leukemia carrying a translocation of this receptor with the TCR, indicating a major role for Notch pathway in T-cell regulation. Many studies have demonstrated that early stages of the T-cell development require Notch1 activation (Wilson et al. 2001); (Radtke et al. 1999), which is achieved by the interaction of the receptor expressed in the precursor cells with Dll4 ligand expressed by the thymic stroma (Hozumi et al. 2008; Koch et al. 2008). During T-cell development, early thymic progenitors (ETP) CD3-CD44+cKit+ colonize the thymus and following Notch signaling activation progress to T-cell differentiation. Consistently, alterations in this pathway has devastating effects on the homeostasis of the T-cell compartment, and in fact Notch has been found to be mutated in about 50 % of samples from human T-ALL (Weng et al. 2004). Moreover, ectopic expression of Notch1-IC in bone marrow murine cells results in T-ALL leukemia, and this is a model widely used in different laboratories [reviewed in (Aster et al. 2011)]. After discovering the importance of Notch activation in T-ALL, many efforts have been done to uncover the downstream mechanisms responsible for the pathogenesis. Several of them have been identified including Myc (Weng et al. 2006), PTEN (Palomero et al. 2007), HES-1 (Espinosa et al. 2010) that are involved in activation of important pathways such as NFkB or PI3K.

Progression from ETP to the mature T-cells can be recapitulated in vitro by exposing the cells to Dll1 or Dll4 ligands (Besseyrias et al. 2007; Schmitt and Zuniga-Pflucker 2002). In vitro, production of T- or B-cells from LSK progenitors can be modulated by Notch as Notch inhibition promotes the B-cell lineage. Similar effects can be achieved by modulating the activity of fringe glycosyl-transferases that enhance Notch-Delta-like interactions (Koch et al. 2001). All these observations were originally interpreted as the effects of Notch in the regulation of a common lymphoid progenitor (CLP) with B- and T-cell potential. However, the finding that ETPs also preserve myeloid potential demands a new and likely more complex interpretation of the model (Kawamoto and Katsura 2009). In this sense, several evidences indicate that Notch negatively regulates myelopoiesis both in vivo and in vitro (Kawamata et al. 2002; Stier et al. 2002), and excessive granulopoiesis has been observed in mice with impaired fucosylation, which precludes Notch activity (Yao et al. 2011). FACS analyses of bone marrow from these animals suggested that Notch is involved in the generation of Myelomonocytic (GMP) progenitors from their Common Myeloid Progenitors (CMP) precursors; similarly, an increase in the GMP but not in the CMP population was observed in mice deficient for Nicastrin (required for γ-secretase activity and Notch processing), that develop a myeloproliferative syndrome, resembling CMML. Interestingly, mutations in the Nicastrin gene have also been found in samples from CMML patients. In addition to the effect of Notch on myeloid progenitors, DLL1-mediated Notch signaling promotes Dendritic Cell (DC) differentiation at expense of the myeloid lineage in vitro (Ohishi et al. 2001; Olivier et al. 2006), although in vivo studies indicated that more than a requirement; Notch signaling can modulate the production of specific DC populations (Caton et al. 2007). Finally, whole genome sequencing approach has revealed the presence of Notch1 mutations in malignancies such as B-chronic lymphocytic leukemia (B-CLL) (Puente et al. 2011). This is an unexpected finding since Notch2 rather than Notch1 regulates B-cell differentiation.

To our understanding, the great variety of phenotypes that are observed following Notch manipulation suggest that this is not just a Notch ON/OFF issue, but different degrees of Notch activity exert specific effects in particular hematopoietic lineages (Bigas and Espinosa 2012).

3.5 Leukemia Initiating Cells: A Role for Notch?

Increasing evidence supports the hypothesis that a small population of cells resembling stem cells maintains tumors. This concept was experimentally demonstrated in leukemic malignances by Dick and colleagues by showing that only few cells from human Acute Myeloid Leukemia samples can recapitulate the entire disease when transplanted into SCID mice (Lapidot et al. 1994). These cells, that were called Leukemia Initiating Cells (LIC), are slow cycling and as a consequence more resistant to conventional anti-cancer therapies often targeting dividing cells (Dick 2008). LIC also has the capacity to self-renew as well as to produce more proliferative precursors by asymmetric cell division, one of the main characteristics of normal stem cells. However, it has been demonstrated that normal HSC can divide both symmetrically and asymmetrically in a balance that is influenced by internal and external signals, and specific oncogenic fusion proteins can disrupt this balance (Wu et al. 2007). In a previous section, we have listed some evidences that associate Notch with the asymmetric division in normal HSCs. Thus, it is tempting to speculate that altered Notch activity in leukemic cells may influence the capacity of asymmetric cell division that is a required characteristic of the LIC population. In agreement with this possibility, Notch-IC transduced bone marrow progenitors acquired the capacity to grow in serial replating assays (D'Altri et al. 2011), which is an in vitro measure of LIC potential (Lavau et al. 1997). Most importantly, Notch is required for LIC maintenance in human T-ALL cells transplanted into NOD-SCID mice (Armstrong et al. 2009) and in a mouse model of T-ALL as indicated the increased animal survival in serial transplantation assays (Tatarek et al. 2011). However, the exact mechanistic link between Notch function and LIC generation or maintenance has not been yet identified. Therapeutic approaches for T-ALL treatment will benefit from our better understanding of this association.

4 Conclusions

During embryonic development, hematopoietic cells are generated from endothelial precursors, but only a few of them (known as HSCs) will acquire the self-renewal capacity without loosing the hematopoietic identity, a process that depends on Notch. Hematopoietic stemness is maintained during the HSC amplification that occurs in the fetal liver, a process likely involving symmetric division. Whether Notch functions at this stage of hematopoietic development is unknown. Subsequently, HSCs colonize the bone marrow and generate all the blood lineages throughout life while maintaining the stem-cell pool by asymmetric division. Although data in other systems identifies Notch as a main player in stem-cell maintenance, current data indicate that Notch activity is dispensable for this function in the hematopoietic system. However, the identification of new Notch modulators and evaluation of different types of Notch-mutant animals may revisit the role of Notch in the adult HSC (see model in Fig. 1).

Acknowledgments We sincerely apologize to those whose work could not be cited due to space limitation. We thank the members of the Bigas-Espinosa lab for critical discussions. The laboratory is funded by Ministerio Ciencia e Innovación (SAF2007-60080, PLE2009-0111, SAF2010-15450, PI10/01128), Red Temática de Investigación Cooperativa en Cáncer (RTICC) (RD06/0020/0098), AGAUR (2009SGR-23 and CONES2010-0006). LE is an investigator of ISCIII program (02/30279).

References

Armstrong F, Brunet de la Grange P, Gerby B, Rouyez MC, Calvo J, Fontenay M, Boissel N, Dombret H, Baruchel A, Landman-Parker J et al (2009) NOTCH is a key regulator of human T-cell acute leukemia initiating cell activity. Blood 113:1730–1740

Aster JC, Blacklow SC, Pear WS (2011) Notch signalling in T-cell lymphoblastic leukaemia/lymphoma and other haematological malignancies. J Pathol 223:262–273

Barker JE (1968) Development of the mouse hematopoietic system. I. Types of hemoglobin produced in embryonic yolk sac and liver. Dev Biol 18:14–29

Beckmann J, Scheitza S, Wernet P, Fischer JC, Giebel B (2007) Asymmetric cell division within the human hematopoietic stem and progenitor cell compartment: identification of asymmetrically segregating proteins. Blood 109:5494–5501

Bertrand JY, Giroux S, Golub R, Klaine M, Jalil A, Boucontet L, Godin I, Cumano A (2005) Characterization of purified intraembryonic hematopoietic stem cells as a tool to define their site of origin. Proc Natl Acad Sci U S A 102:134–139

Bertrand JY, Cisson JL, Stachura DL, Traver D (2010) Notch signaling distinguishes two waves of definitive hematopoiesis in the zebrafish embryo. Blood 115:2777–2783

Besseyrias V, Fiorini E, Strobl LJ, Zimber-Strobl U, Dumortier A, Koch U, Arcangeli ML, Ezine S, Macdonald HR, Radtke F (2007) Hierarchy of Notch-Delta interactions promoting T cell lineage commitment and maturation. J Exp Med 204:331–343

Bigas A, Espinosa L (2012) Hematopoietic stem cells: to be or Notch to be. Blood 119(14): 3226–3235

Bigas A, Robert-Moreno A, Espinosa L (2010) The Notch pathway in the developing hematopoietic system. Int J Dev Biol 54:1175–1188

Burns CE, Traver D, Mayhall E, Shepard JL, Zon LI (2005) Hematopoietic stem cell fate is established by the Notch-Runx pathway. Genes Dev 19:2331–2342

Calvi LM, Adams GB, Weibrecht KW, Weber JM, Olson DP, Knight MC, Martin RP, Schipani E, Divieti P, Bringhurst FR et al (2003) Osteoblastic cells regulate the haematopoietic stem cell niche. Nature 425:841–846

Carlesso N, Aster JC, Sklar J, Scadden DT (1999) Notch1-induced delay of human hematopoietic progenitor cell differentiation is associated with altered cell cycle kinetics. Blood 93:838–848

Caton ML, Smith-Raska MR, Reizis B (2007) Notch-RBP-J signaling controls the homeostasis of CD8- dendritic cells in the spleen. J Exp Med 204:1653–1664

Chen VC, Stull R, Joo D, Cheng X, Keller G (2008) Notch signaling respecifies the hemangioblast to a cardiac fate. Nat Biotechnol 26:1169–1178

Chen MJ, Yokomizo T, Zeigler BM, Dzierzak E, Speck NA (2009) Runx1 is required for the endothelial to haematopoietic cell transition but not thereafter. Nature 457:887–891

Cheng X, Huber TL, Chen VC, Gadue P, Keller GM (2008) Numb mediates the interaction between Wnt and Notch to modulate primitive erythropoietic specification from the hemangioblast. Development 135:3447–3458

Choi K, Kennedy M, Kazarov A, Papadimitriou JC, Keller G (1998) A common precursor for hematopoietic and endothelial cells. Development 125:725–732

Ciau-Uitz A, Walmsley M, Patient R (2000) Distinct origins of adult and embryonic blood in Xenopus. Cell 102:787–796

Conboy IM, Rando TA (2002) The regulation of Notch signaling controls satellite cell activation and cell fate determination in postnatal myogenesis. Dev Cell 3:397–409

Couturier L, Vodovar N, Schweisguth F (2012) Endocytosis by Numb breaks Notch symmetry at cytokinesis. Nat Cell Biol 14:131–139

Cumano A, Godin I (2007) Ontogeny of the hematopoietic system. Annu Rev Immunol 25: 745–785

D'Altri T, Gonzalez J, Aifantis I, Espinosa L, Bigas A (2011) Hes1 expression and CYLD repression are essential events downstream of Notch1 in T-cell leukemia. Cell Cycle 10:1031–1036

de Andres-Aguayo L, Varas F, Kallin EM, Infante JF, Wurst W, Floss T, Graf T (2011) Musashi 2 is a regulator of the HSC compartment identified by a retroviral insertion screen and knockout mice. Blood 118:554–564
de Bruijn MF, Speck NA, Peeters MC, Dzierzak E (2000) Definitive hematopoietic stem cells first develop within the major arterial regions of the mouse embryo. EMBO J 19:2465–2474
de Bruijn MF, Ma X, Robin C, Ottersbach K, Sanchez MJ, Dzierzak E (2002) Hematopoietic stem cells localize to the endothelial cell layer in the midgestation mouse aorta. Immunity 16:673–683
Delaney C, Heimfeld S, Brashem-Stein C, Voorhies H, Manger RL, Bernstein ID (2010) Notch-mediated expansion of human cord blood progenitor cells capable of rapid myeloid reconstitution. Nat Med 16:232–236
Dick JE (2008) Stem cell concepts renew cancer research. Blood 112:4793–4807
Dieterlen-Lievre F, Beaupain D, Martin C (1976) Origin of erythropoietic stem cells in avian development: shift from the yolk sac to an intraembryonic site. Ann Immunol (Paris) 127:857–863
Duarte A, Hirashima M, Benedito R, Trindade A, Diniz P, Bekman E, Costa L, Henrique D, Rossant J (2004) Dosage-sensitive requirement for mouse Dll4 in artery development. Genes Dev 18:2474–2478
Dzierzak E, Speck NA (2008) Of lineage and legacy: the development of mammalian hematopoietic stem cells. Nat Immunol 9:129–136
Eilken HM, Nishikawa S, Schroeder T (2009) Continuous single-cell imaging of blood generation from haemogenic endothelium. Nature 457:896–900
Espinosa L, Cathelin S, D'Altri T, Trimarchi T, Statnikov A, Guiu J, Rodilla V, Ingles-Esteve J, Nomdedeu J, Bellosillo B et al (2010) The Notch/Hes1 pathway sustains NF-kappaB activation through CYLD repression in T cell leukemia. Cancer Cell 18:268–281
Ferkowicz MJ, Yoder MC (2005) Blood island formation: longstanding observations and modern interpretations. Exp Hematol 33:1041–1047
Fraser ST, Ogawa M, Yu RT, Nishikawa S, Yoder MC (2002) Definitive hematopoietic commitment within the embryonic vascular endothelial-cadherin(+) population. Exp Hematol 30:1070–1078
Furthauer M, Gonzalez-Gaitan M (2009) Endocytosis, asymmetric cell division, stem cells and cancer: unus pro omnibus, omnes pro uno. Mol Oncol 3:339–353
Gekas C, Dieterlen-Lievre F, Orkin SH, Mikkola HK (2005) The placenta is a niche for hematopoietic stem cells. Dev Cell 8:365–375
Giebel B, Zhang T, Beckmann J, Spanholtz J, Wernet P, Ho AD, Punzel M (2006) Primitive human hematopoietic cells give rise to differentially specified daughter cells upon their initial cell division. Blood 107:2146–2152
Hadland BK, Huppert SS, Kanungo J, Xue Y, Jiang R, Gridley T, Conlon RA, Cheng AM, Kopan R, Longmore GD (2004) A requirement for Notch1 distinguishes 2 phases of definitive hematopoiesis during development. Blood 104:3097–3105
Hozumi K, Mailhos C, Negishi N, Hirano K, Yahata T, Ando K, Zuklys S, Hollander GA, Shima DT, Habu S (2008) Delta-like 4 is indispensable in thymic environment specific for T cell development. J Exp Med 205:2507–2513
Irion S, Clarke RL, Luche H, Kim I, Morrison SJ, Fehling HJ, Keller GM (2010) Temporal specification of blood progenitors from mouse embryonic stem cells and induced pluripotent stem cells. Development 137:2829–2839
Itoh M, Kim CH, Palardy G, Oda T, Jiang YJ, Maust D, Yeo SY, Lorick K, Wright GJ, Ariza-McNaughton L et al (2003) Mind bomb is a ubiquitin ligase that is essential for efficient activation of Notch signaling by Delta. Dev Cell 4:67–82
Ivanovs A, Rybtsov S, Welch L, Anderson RA, Turner ML, Medvinsky A (2011) Highly potent human hematopoietic stem cells first emerge in the intraembryonic aorta-gonad-mesonephros region. J Exp Med 208:2417–2427
Kageyama R, Ohtsuka T, Kobayashi T (2007) The Hes gene family: repressors and oscillators that orchestrate embryogenesis. Development 134:1243–1251

Kawamata S, Du C, Li K, Lavau C (2002) Overexpression of the Notch target genes Hes in vivo induces lymphoid and myeloid alterations. Oncogene 21:3855–3863

Kawamoto H, Katsura Y (2009) A new paradigm for hematopoietic cell lineages: revision of the classical concept of the myeloid-lymphoid dichotomy. Trends Immunol 30:193–200

Kharas MG, Lengner CJ, Al-Shahrour F, Bullinger L, Ball B, Zaidi S, Morgan K, Tam W, Paktinat M, Okabe R et al (2010) Musashi-2 regulates normal hematopoiesis and promotes aggressive myeloid leukemia. Nat Med 16:903–908

Kim YW, Koo BK, Jeong HW, Yoon MJ, Song R, Shin J, Jeong DC, Kim SH, Kong YY (2008) Defective Notch activation in microenvironment leads to myeloproliferative disease. Blood 112:4628–4638

Koch U, Lacombe TA, Holland D, Bowman JL, Cohen BL, Egan SE, Guidos CJ (2001) Subversion of the T/B lineage decision in the thymus by lunatic fringe-mediated inhibition of Notch-1. Immunity 15:225–236

Koch U, Fiorini E, Benedito R, Besseyrias V, Schuster-Gossler K, Pierres M, Manley NR, Duarte A, Macdonald HR, Radtke F (2008) Delta-like 4 is the essential, nonredundant ligand for Notch1 during thymic T cell lineage commitment. J Exp Med 205:2515–2523

Koo BK, Lim HS, Song R, Yoon MJ, Yoon KJ, Moon JS, Kim YW, Kwon MC, Yoo KW, Kong MP et al (2005) Mind bomb 1 is essential for generating functional Notch ligands to activate Notch. Development 132:3459–3470

Kopan R, Ilagan MX (2009) The canonical Notch signaling pathway: unfolding the activation mechanism. Cell 137:216–233

Krebs LT, Shutter JR, Tanigaki K, Honjo T, Stark KL, Gridley T (2004) Haploinsufficient lethality and formation of arteriovenous malformations in Notch pathway mutants. Genes Dev 18:2469–2473

Kumano K, Chiba S, Kunisato A, Sata M, Saito T, Nakagami-Yamaguchi E, Yamaguchi T, Masuda S, Shimizu K, Takahashi T et al (2003) Notch1 but not Notch2 is essential for generating hematopoietic stem cells from endothelial cells. Immunity 18:699–711

Kyba M, Perlingeiro RC, Daley GQ (2002) HoxB4 confers definitive lymphoid-myeloid engraftment potential on embryonic stem cell and yolk sac hematopoietic progenitors. Cell 109:29–37

Lancrin C, Sroczynska P, Stephenson C, Allen T, Kouskoff V, Lacaud G (2009) The haemangioblast generates haematopoietic cells through a haemogenic endothelium stage. Nature 457:892–895

Lapidot T, Sirard C, Vormoor J, Murdoch B, Hoang T, Caceres-Cortes J, Minden M, Paterson B, Caligiuri MA, Dick JE (1994) A cell initiating human acute myeloid leukaemia after transplantation into SCID mice. Nature 367:645–648

Lavau C, Szilvassy SJ, Slany R, Cleary ML (1997) Immortalization and leukemic transformation of a myelomonocytic precursor by retrovirally transduced HRX-ENL. EMBO J 16:4226–4237

Lawson ND, Scheer N, Pham VN, Kim CH, Chitnis AB, Campos-Ortega JA, Weinstein BM (2001) Notch signaling is required for arterial-venous differentiation during embryonic vascular development. Development 128:3675–3683

Le Borgne R, Remaud S, Hamel S, Schweisguth F (2005) Two distinct E3 ubiquitin ligases have complementary functions in the regulation of delta and serrate signaling in Drosophila. PLoS Biol 3:e96

Ledran MH, Krassowska A, Armstrong L, Dimmick I, Renstrom J, Lang R, Yung S, Santibanez-Coref M, Dzierzak E, Stojkovic M et al (2008) Efficient hematopoietic differentiation of human embryonic stem cells on stromal cells derived from hematopoietic niches. Cell Stem Cell 3:85–98

Maillard I, Koch U, Dumortier A, Shestova O, Xu L, Sai H, Pross SE, Aster JC, Bhandoola A, Radtke F et al (2008) Canonical notch signaling is dispensable for the maintenance of adult hematopoietic stem cells. Cell Stem Cell 2:356–366

Mancini SJ, Mantei N, Dumortier A, Suter U, MacDonald HR, Radtke F (2005) Jagged1-dependent Notch signaling is dispensable for hematopoietic stem cell self-renewal and differentiation. Blood 105:2340–2342

McCulloch EA, Till JE (1960) The radiation sensitivity of normal mouse bone marrow cells, determined by quantitative marrow transplantation into irradiated mice. Radiat Res 13:115–125

McGrath KE, Frame JM, Fromm GJ, Koniski AD, Kingsley PD, Little J, Bulger M, Palis J (2011) A transient definitive erythroid lineage with unique regulation of the beta-globin locus in the mammalian embryo. Blood 117:4600–4608

Muller AM, Medvinsky A, Strouboulis J, Grosveld F, Dzierzak E (1994) Development of hematopoietic stem cell activity in the mouse embryo. Immunity 1:291–301

Nostro MC, Cheng X, Keller GM, Gadue P (2008) Wnt, activin, and BMP signaling regulate distinct stages in the developmental pathway from embryonic stem cells to blood. Cell Stem Cell 2:60–71

Oberlin E, Fleury M, Clay D, Petit-Cocault L, Candelier JJ, Mennesson B, Jaffredo T, Souyri M (2010) VE-cadherin expression allows identification of a new class of hematopoietic stem cells within human embryonic liver. Blood 116:4444–4455

Ohishi K, Varnum-Finney B, Serda RE, Anasetti C, Bernstein ID (2001) The Notch ligand, Delta-1, inhibits the differentiation of monocytes into macrophages but permits their differentiation into dendritic cells. Blood 98:1402–1407

Ohlstein B, Spradling A (2007) Multipotent Drosophila intestinal stem cells specify daughter cell fates by differential notch signaling. Science 315:988–992

Oka C, Nakano T, Wakeham A, de la Pompa JL, Mori C, Sakai T, Okazaki S, Kawaichi M, Shiota K, Mak TW et al (1995) Disruption of the mouse RBP-J kappa gene results in early embryonic death. Development 121:3291–3301

Olivier A, Lauret E, Gonin P, Galy A (2006) The Notch ligand delta-1 is a hematopoietic development cofactor for plasmacytoid dendritic cells. Blood 107:2694–2701

Ottersbach K, Dzierzak E (2005) The murine placenta contains hematopoietic stem cells within the vascular labyrinth region. Dev Cell 8:377–387

Palis J, Robertson S, Kennedy M, Wall C, Keller G (1999) Development of erythroid and myeloid progenitors in the yolk sac and embryo proper of the mouse. Development 126:5073–5084

Palomero T, Sulis ML, Cortina M, Real PJ, Barnes K, Ciofani M, Caparros E, Buteau J, Brown K, Perkins SL et al (2007) Mutational loss of PTEN induces resistance to NOTCH1 inhibition in T-cell leukemia. Nat Med 13:1203–1210

Puente XS, Pinyol M, Quesada V, Conde L, Ordonez GR, Villamor N, Escaramis G, Jares P, Bea S, Gonzalez-Diaz M et al (2011) Whole-genome sequencing identifies recurrent mutations in chronic lymphocytic leukaemia. Nature 475:101–105

Radtke F, Wilson A, Stark G, Bauer M, van Meerwijk J, MacDonald HR, Aguet M (1999) Deficient T cell fate specification in mice with an induced inactivation of Notch1. Immunity 10:547–558

Robert-Moreno A, Espinosa L, de la Pompa JL, Bigas A (2005) RBPjkappa-dependent Notch function regulates Gata2 and is essential for the formation of intra-embryonic hematopoietic cells. Development 132:1117–1126

Robert-Moreno A, Espinosa L, Sanchez MJ, de la Pompa JL, Bigas A (2007) The notch pathway positively regulates programmed cell death during erythroid differentiation. Leukemia 21:1496–1503

Robert-Moreno A, Guiu J, Ruiz-Herguido C, Lopez ME, Ingles-Esteve J, Riera L, Tipping A, Enver T, Dzierzak E, Gridley T et al (2008) Impaired embryonic haematopoiesis yet normal arterial development in the absence of the Notch ligand Jagged1. EMBO J 27:1886–1895

Rybtsov S, Sobiesiak M, Taoudi S, Souilhol C, Senserrich J, Liakhovitskaia A, Ivanovs A, Frampton J, Zhao S, Medvinsky A (2011) Hierarchical organization and early hematopoietic specification of the developing HSC lineage in the AGM region. J Exp Med 208:1305–1315

Schmitt TM, Zuniga-Pflucker JC (2002) Induction of T cell development from hematopoietic progenitor cells by delta-like-1 in vitro. Immunity 17:749–756

Schroeder T, Fraser ST, Ogawa M, Nishikawa S, Oka C, Bornkamm GW, Honjo T, Just U (2003) Recombination signal sequence-binding protein Jkappa alters mesodermal cell fate decisions by suppressing cardiomyogenesis. Proc Natl Acad Sci U S A 100:4018–4023

Sorensen LK, Brooke BS, Li DY, Urness LD (2003) Loss of distinct arterial and venous boundaries in mice lacking endoglin, a vascular-specific TGFbeta coreceptor. Dev Biol 261:235–250

Stier S, Cheng T, Dombkowski D, Carlesso N, Scadden DT (2002) Notch1 activation increases hematopoietic stem cell self-renewal in vivo and favors lymphoid over myeloid lineage outcome. Blood 99:2369–2378

Tatarek J, Cullion K, Ashworth T, Gerstein R, Aster JC, Kelliher MA (2011) Notch1 inhibition targets the leukemia-initiating cells in a Tal1/Lmo2 mouse model of T-ALL. Blood 118:1579–1590

Urness LD, Sorensen LK, Li DY (2000) Arteriovenous malformations in mice lacking activin receptor-like kinase-1. Nat Genet 26:328–331

Varnum-Finney B, Xu L, Brashem-Stein C, Nourigat C, Flowers D, Bakkour S, Pear WS, Bernstein ID (2000) Pluripotent, cytokine-dependent, hematopoietic stem cells are immortalized by constitutive notch1 signaling. Nat Med 6:1278–1281

Varnum-Finney B, Brashem-Stein C, Bernstein ID (2003) Combined effects of Notch signaling and cytokines induce a multiple log increase in precursors with lymphoid and myeloid reconstituting ability. Blood 101:1784–1789

Varnum-Finney B, Halasz LM, Sun M, Gridley T, Radtke F, Bernstein ID (2011) Notch2 governs the rate of generation of mouse long- and short-term repopulating stem cells. J Clin Invest 121:1207–1216

Walmsley M, Ciau-Uitz A, Patient R (2002) Adult and embryonic blood and endothelium derive from distinct precursor populations which are differentially programmed by BMP in Xenopus. Development 129:5683–5695

Wang MM (2011) Notch signaling and Notch signaling modifiers. Int J Biochem Cell Biol 43:1550–1562

Wang Y, Shao L, Shi S, Harris RJ, Spellman MW, Stanley P, Haltiwanger RS (2001) Modification of epidermal growth factor-like repeats with O-fucose. Molecular cloning and expression of a novel GDP-fucose protein O-fucosyltransferase. J Biol Chem 276:40338–40345

Wang L, Menendez P, Shojaei F, Li L, Mazurier F, Dick JE, Cerdan C, Levac K, Bhatia M (2005a) Generation of hematopoietic repopulating cells from human embryonic stem cells independent of ectopic HOXB4 expression. J Exp Med 201:1603–1614

Wang Y, Yates F, Naveiras O, Ernst P, Daley GQ (2005b) Embryonic stem cell-derived hematopoietic stem cells. Proc Natl Acad Sci U S A 102:19081–19086

Weng AP, Ferrando AA, Lee W, Morris JPt, Silverman LB, Sanchez-Irizarry C, Blacklow SC, Look AT, Aster JC (2004) Activating mutations of NOTCH1 in human T cell acute lymphoblastic leukemia. Science 306:269–271

Weng AP, Millholland JM, Yashiro-Ohtani Y, Arcangeli ML, Lau A, Wai C, Del Bianco C, Rodriguez CG, Sai H, Tobias J et al (2006) c-Myc is an important direct target of Notch1 in T-cell acute lymphoblastic leukemia/lymphoma. Genes Dev 20:2096–2109

Wilson A, MacDonald HR, Radtke F (2001) Notch 1-deficient common lymphoid precursors adopt a B cell fate in the thymus. J Exp Med 194:1003–1012

Wu M, Kwon HY, Rattis F, Blum J, Zhao C, Ashkenazi R, Jackson TL, Gaiano N, Oliver T, Reya T (2007) Imaging hematopoietic precursor division in real time. Cell Stem Cell 1:541–554

Yamamoto M, Morita R, Mizoguchi T, Matsuo H, Isoda M, Ishitani T, Chitnis AB, Matsumoto K, Crump JG, Hozumi K et al (2010) Mib-Jag1-Notch signalling regulates patterning and structural roles of the notochord by controlling cell-fate decisions. Development 137:2527–2537

Yang LT, Nichols JT, Yao C, Manilay JO, Robey EA, Weinmaster G (2005) Fringe glycosyltransferases differentially modulate Notch1 proteolysis induced by Delta1 and Jagged1. Mol Biol Cell 16:927–942

Yao D, Huang Y, Huang X, Wang W, Yan Q, Wei L, Xin W, Gerson S, Stanley P, Lowe JB et al (2011) Protein O-fucosyltransferase 1 (Pofut1) regulates lymphoid and myeloid homeostasis through modulation of Notch receptor ligand interactions. Blood 117:5652–5662

You LR, Lin FJ, Lee CT, DeMayo FJ, Tsai MJ, Tsai SY (2005) Suppression of Notch signalling by the COUP-TFII transcription factor regulates vein identity. Nature 435:98–104

Yu X, Zou J, Ye Z, Hammond H, Chen G, Tokunaga A, Mali P, Li YM, Civin C, Gaiano N et al (2008) Notch signaling activation in human embryonic stem cells is required for embryonic, but not trophoblastic, lineage commitment. Cell Stem Cell 2:461–471

Zambidis ET, Peault B, Park TS, Bunz F, Civin CI (2005) Hematopoietic differentiation of human embryonic stem cells progresses through sequential hematoendothelial, primitive, and definitive stages resembling human yolk sac development. Blood 106:860–870

Zhong W, Feder JN, Jiang MM, Jan LY, Jan YN (1996) Asymmetric localization of a mammalian numb homolog during mouse cortical neurogenesis. Neuron 17:43–53

Zovein AC, Hofmann JJ, Lynch M, French WJ, Turlo KA, Yang Y, Becker MS, Zanetta L, Dejana E, Gasson JC et al (2008) Fate tracing reveals the endothelial origin of hematopoietic stem cells. Cell Stem Cell 3:625–636

Notch Receptor-Ligand Interactions During T Cell Development, a Ligand Endocytosis-Driven Mechanism

Divya K. Shah and Juan Carlos Zúñiga-Pflücker

Abstract Notch signaling plays an important role during the development of different cell types and tissues. The role of Notch signaling in lymphocyte development, in particular in the development and commitment to the T cell lineage, has been the focus of research for many years. Notch signaling is absolutely required during the commitment and early stages of T cell development. Activation of the Notch signaling pathway is initiated by ligand-receptor interactions and appears to require active endocytosis of Notch ligands. Studies addressing the mechanism underlying endocytosis of Notch ligands have helped to identify the main players important and necessary for this process. Here, we review the Notch ligands, and the proposed models of Notch activation by Notch ligand endocytosis, highlighting key molecules involved. In particular, we discuss recent studies on Notch ligands involved in T cell development, current studies aimed at elucidating the relevance of Notch ligand endocytosis during T cell development and the identification of key players necessary for ligand endocytosis in the thymus and during T cell development.

Abbreviations

ADAM	A disintegrin and metalloprotease
ANK	Ankyrin
BM	Bone marrow
CD	Clusters of differentiation
CDK	Cyclin dependent kinase

D. K. Shah (✉) · J. C. Zúñiga-Pflücker
Department of Immunology, Sunnybrook Research Institute, University of Toronto, 2075 Bayview Avenue, Toronto, ON M4 N 3M5, Canada
e-mail: dkshah@sri.utoronto.ca

J. C. Zúñiga-Pflücker
e-mail: jczp@sri.utoronto.ca

Current Topics in Microbiology and Immunology (2012) 360: 19–46
DOI: 10.1007/82_2012_225

Published Online: 15 May 2012

CSL	CBF-1 in humans, suppressor of hairless in *Drosophila*, lag in *C. elegans*
Dll	Delta-like
DN	Double negative
DOS	Delta and OSM-11-like proteins
DP	Double positive
DSL	Delta/Serrate/Lag2
EGF	Epidermal growth factor
ETP	Early thymic progenitor
FL	Fetal liver
FoxN1	Forkhead box protein N1
GPI	Glycosylphosphatidylinositol
GSL	Glycosphingolipid
HSC	Hematopoietic stem cell
ICN	Intracellular Notch
ISP	Immature single positive
Jag	Jagged
LDL	Low density lipoprotein
Lfng	Lunatic fringe
Lqf	Lquid facets
LNR	LIN-12 Notch repeats
MamL	Mastermind-Like
Mib	Mindbomb
MINT	Msx interacting nuclear target protein
mRNA	Messenger ribonuclease
MMTV	Mouse mammary tumor virus
NECD	Notch extracellular domain
Neur	Neuralized
NHR	Neuralized homology repeats
NLS	Nuclear localization signal
NRR	Negative regulatory region
PEST	Polypeptide enriched in proline (P), glutamic acid (E), serine (S), and threonine (T)
PDZL	PSD-95/Dlg/ZO-1-ligand
PIPs	Phosphoinositol phosphates
RAG	Recombination activating gene
RAM	RBPJ association module
RBPJ	Recombination binding protein joining-kappa
RING	Really interesting new gene
siRNA	Small interference ribonuclease
SMRT	Silencing mediator of retinoid and thyroid receptor
SP	Single positive
TAD	Transactivation domain
TCR	T cell receptor

TEC	Thymic epithelial cells
TF	Transcription factor
TSPs	Thymic seeding progenitors
WT	Wildtype
Ub	Ubiquitin

Contents

1 T Cell Development

The thymus is the site of development of the majority of T cells, and provides an essential stromal microenvironment necessary for the generation of a functional and diverse T cell repertoire, which is self-restricted and self-tolerant (Petrie and Zuniga-Pflucker 2007).

Cells emigrating to the thymus from either the fetal liver or the adult bone marrow, undergo a series of well-characterized checkpoints, and become increasingly more restricted to the T cell lineage, while eliminating the possibility of other cell fates. These emigrating thymic seeding progenitor cells (TSPs) have been the focus of many studies and is reviewed in Bhandoola et al. (2007).

Different stages of T cell development have been classically defined on the basis of CD4 and CD8 expression. The starting progenitors within the thymus lack the expression of CD4 and CD8, and are termed double negative (DN) cells. The DN subset can be further subdivided by the expression of the cell surface markers CD44 and CD25, with the following developmental sequence: DN1

($CD44^+$ $CD25^-$); DN2 ($CD44^+$ $CD25^+$); DN3 ($CD44^-$ $CD25^+$); and, DN4 ($CD44^-$ $CD25^-$). By the DN3 stage, thymocytes are fully committed to the T cell lineage and their further differentiation requires the expression of the pre-T cell receptor (TCR), a critical developmental checkpoint termed β-selection (Michie and Zuniga-Pflucker 2002). This results in the generation of CD4 and CD8 (double positive, DP) expressing cells, which rearrange their TCR-α gene loci, to produce an $\alpha\beta$-TCR, which following interactions with major histocompatibility complex (MHC)-class-I or MHC-class-II molecules may lead to the selection of $CD8^+$ or $CD4^+$ single positive (SP) cells (Ciofani and Zuniga-Pflucker 2007; Petrie and Zuniga-Pflucker 2007).

$\gamma\delta$ T cells are able to develop from the DN1-DN3 stages, and do not undergo selection checkpoints in the same manner as $\alpha\beta$ developing thymocytes. Instead it seems that the strength of $\gamma\delta$-TCR signals directs the adoption of this lineage outcome (Hayes et al. 2005; Wong and Zuniga-Pflucker 2010).

There are many factors that are involved in the process of thymocyte differentiation, including cytokines, transcription factors, and genes that are important in determining cell fate and patterning in the developing embryo, such as Hh, BMPs, and Notch. All these factors act in conjunction with each other and their interplay is important in the production of a successful pool of T cells that exit the thymus. The reader is referred to recent reviews (Ciofani and Zuniga-Pflucker 2007; Thompson and Zuniga-Pflucker 2011; Varas et al. 2003). The focus of this review will be the role of components of the Notch signaling pathway in early T cell development, in particular concentrating on Notch ligands and Notch ligand endocytosis, and recent advances in this field in relation to T cell development.

2 The Notch Signaling Pathway

Notch was first identified in *Drosophila,* after the observation of a heterozygous loss-of function mutation that exhibited a 'notched' wing margin phenotype (Morgan 1917). Notch was cloned in 1983 (Artavanis-Tsakonas et al. 1983) and since then a plethora of work has identified that it is highly conserved in many species. The Notch receptor is activated by Delta-Serrate-Lag2 (DSL) ligands, of which there are two families: Delta and Serrate (Bray 2006; Tien et al. 2009). In mammals, there are four Notch receptors (Notch-1, -2, -3, and -4) (Del Amo et al. 1992; Ellisen et al. 1991; Lardelli et al. 1994; Lardelli and Lendahl 1993; Uyttendaele et al. 1996; Weinmaster et al. 1992), which bind and interact with two Serrate family homologues (Jagged-1 and -2), and three Delta family homologues (Delta-like-1, -3, and -4) (Bettenhausen et al. 1995; Dunwoodie et al. 1997; Lindsell et al. 1995; Shawber et al. 1996; Shutter et al. 2000). As illustrated in Fig. 1, upon ligand-receptor interactions the heterodimeric Notch receptor undergoes an S2 cleavage in its extracellular domain, mediated by a disintegrin and metalloprotease (ADAM). The transmembrane associated Notch is then further cleaved by a γ-secretase multiprotein complex (S3 cleavage) consisting of presenilin, nicastrin, APH-1, and PEN-2 proteins, releasing the active intracellular Notch (ICN) into the cytoplasm (Kopan and Ilagan 2009). The ICN translocates into the nucleus, and interacts with the helix-loop-helix transcription factor CSL (CBF-1 in humans, Suppressor of Hairless in *Drosophila*, Lag in *C. elegans*)/RBPJ (recombination binding protein joining-kappa) in mice, displacing corepressors such as silencing mediator of retinoid and thyroid receptor (SMRT) and Msx interacting nuclear target protein (MINT) (Kao et al. 1998; Kuroda et al. 2003; Oswald et al. 2005), and recruits Mastermind-like (MamL) proteins, which then aid in recruiting coactivators p300 and histone acetyl transferases, thus promoting target gene transcription (Kovall 2007; Nichols et al. 2007; Wu et al. 2000). There are three mammalian homologues of MamL, which are able to bind to all four Notch receptors (Fryer et al. 2002; Wu et al. 2002). Target genes of Notch signaling include: *hairy enhancer of split-1* (*hes-1*), *deltex1*, and tissue-specific targets such as: *pre-Tα* and *cd25* (Deftos et al. 2000; Jarriault et al. 1995; Reizis and Leder 2002).

3 Notch Receptors

In mammals, the Notch receptor is produced as a single protein, which is proteolytically cleaved by a furin-like convertase (S1 cleavage) in the cytoplasm producing a heterodimeric transmembrane receptor, consisting of non-covalently associated extracellular and intracellular domains that are expressed at the cell surface (Blaumueller et al. 1997). The structure of the extracellular domain includes a series of 29–36 epidermal growth factor (EGF)-like repeats, important for ligand binding, in particular EGF repeats 11–12 are important for Notch ligand interactions (Rebay et al. 1991), a negative regulatory region (NRR) which is composed of three cysteine-rich LIN-12 Notch repeats (LNR), important for preventing ligand-independent activation and a heterodimerization domain. The intracellular portion consists of an RBPJ association module (RAM) domain, two nuclear localization signals (NLS), and seven ankyrin repeats (ANK), of which the RAM and the ANK repeats are both involved in binding to CSL/RBPJ and MamL (Tamura et al. 1995). There is also a transactivation domain (TAD) and a polypeptide enriched in proline (P), glutamic acid (E), serine (S), and threonine (T) (PEST) domain, which regulates protein stablilty (Kopan and Ilagan 2009). Proteolytic processing of Notch receptors is carried out by ADAM proteases (S2 cleavage) and γ-secretase activity of the presenilin multiprotein complex (S3 cleavage) (Kopan and Ilagan 2009).

The EGF repeats on the Notch receptor can undergo O linked-glycosylation, O linked-glucose modifications are mediated by the glycosyltransferase Rumi in flies (Acar et al. 2008) and are thought to improve binding of Notch to its ligand. O-fucose modifications are mediated by the fucosyltransferase Pofut in mammals (Ofut in flies) (Okajima and Irvine 2002; Sasamura et al. 2003; Shi and Stanley 2003). Fringe proteins then extend the O-fucose modified EGF repeats further

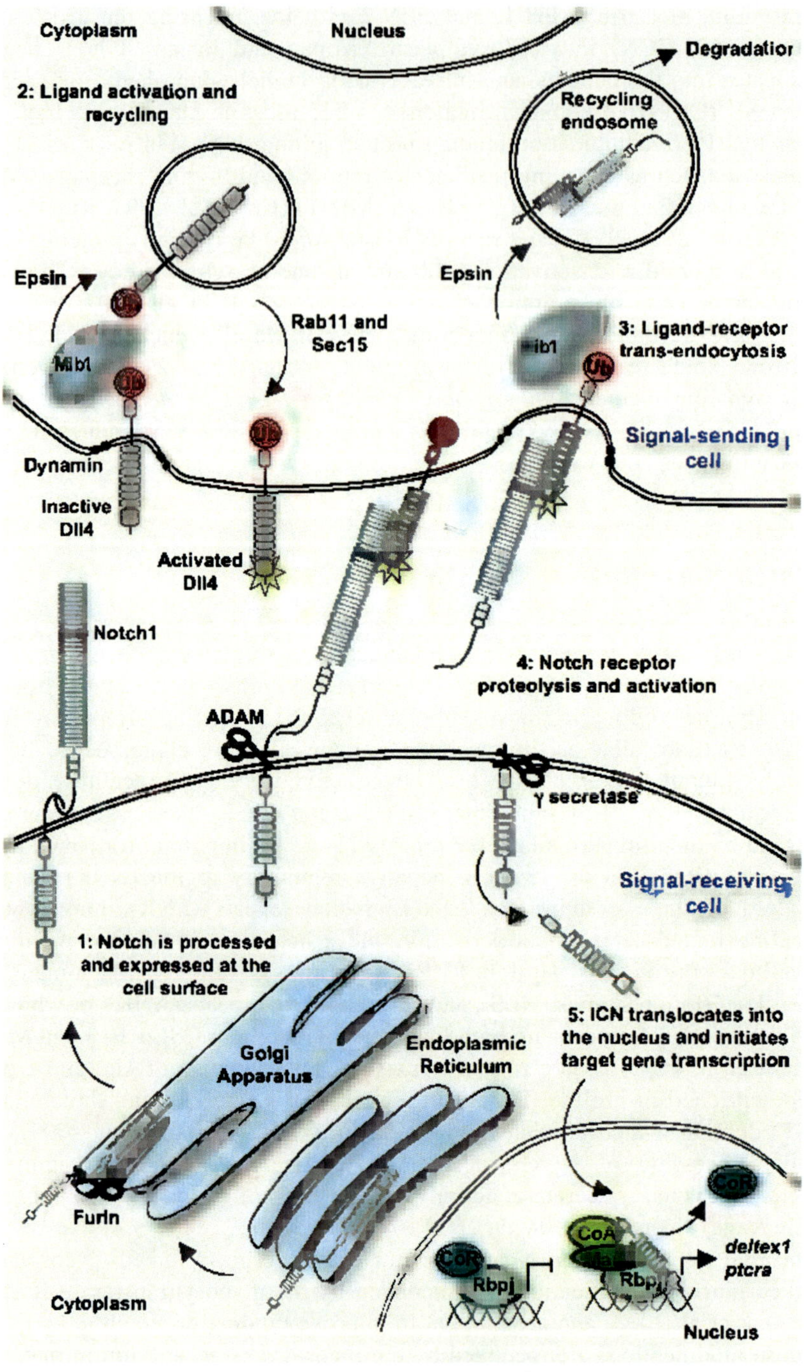
Cytoplasm
Nucleus
Degradation
2: Ligand activation and recycling
Recycling endosome
Epsin
Epsin
Rab11 and Sec15
Mib1
3: Ligand-receptor trans-endocytosis
Signal-sending cell
Dynamin
Inactive Dll4
Activated Dll4
Notch1
4: Notch receptor proteolysis and activation
ADAM
γ secretase
Signal-receiving cell
1: Notch is processed and expressed at the cell surface
5: ICN translocates into the nucleus and initiates target gene transcription
Golgi Apparatus
Endoplasmic Reticulum
Furin
CoR
CoA
Rbpj
deltex1
ptcra
Cytoplasm
Nucleus

◀**Fig. 1** *Schematic of Notch1 receptor activation by Delta-like 4 in the thymus*. Activation of the Notch signaling pathway is initiated by receptor-ligand interactions. The biosynthesis, processing, and activation required for Notch1 and Dll4 interactions are depicted as indicated. (1) The Notch1 receptor is produced as a single protein in the ER that is proteolytically cleaved by a furin-like convertase within the Golgi, generating a heterodimeric receptor that is non-covalently expressed on the cell surface of thymocytes, the 'signal-receiving cells'. Notch1 can also undergo O-linked glycosylation and Fringe modifications in the cytoplasm (not shown). (2) Dll4 expressed by the thymic epithelial cells, the 'signal-sending cells', may require activation as suggested by the 'Recycling model'—in which ligands are produced and expressed in an inactive form on the cell surface, followed by internalization mediated by Mib1 ubiquitinylation, which is thought to promote the "activation" of the ligand prior to its reexpression on the cell surface, via a process that also involves Dynamin, Epsin, Rab11, and Sec15. (3) The DSL domain of Dll4 and the EGF repeats (11–12) on the Notch1 receptor are important for ligand-receptor interactions. The 'Mechanotransduction model' suggests that when Notch1 binds Dll4, the Mib1-ubiquitinylated Dll4 undergoes endocytosis, which "pulls" on Notch1 to induce a conformational change in the Notch receptor that leads to the trans-endocytosis of the Notch ECD into the signal-sending cell and it is then targeted for degradation. (4) Following the removal of Notch ECD by Dll4, an ADAM cleavage site is exposed on the Notch receptor. The ADAM-cleaved Notch receptor becomes internalized (not shown), and then undergoes γ-secretase processing, releasing ICN into the cytoplasm, allowing it to translocate into the nucleus where it binds to RBPJ, displacing corepressors, recruiting MamL and co-activators, and inducing target gene transcription, such as: *deltex1* and *ptrca*

(Panin et al. 1997). There are three mammalian homologues of Fringe: Lunatic Fringe, Manic Fringe, and Radical Fringe which work in a cell autonomous manner, to potentiate signaling by Delta family members and restrict signaling by Serrate family members (Johnston et al. 1997).

4 DSL Notch Ligands

DSL Notch ligands are transmembrane proteins, which directly interact with Notch receptors at their cell surface. As mentioned earlier there are two families of ligands, the delta family, in mammals: Delta-like-1 (Dll1), Dll3, and Dll4, and the Serrate family, in mammals: Jagged1 (Jag1) and Jag2. The extracellular domain of all DSL ligands consist of an N-terminal (NT) domain, followed by a DSL domain and a series of EGF repeats (Parks et al. 2006; Skwarek et al. 2007). The N-terminal region can be further subdivided into two regions, based on its richness in cysteine, the N1 region: is cysteine-rich and the N2 region: is cysteine free (Parks et al. 2006). Jagged ligands have an additional cysteine-rich region not present in Dll ligands (Vitt et al. 2001). Recently Hamel et al., identified a conserved glycosphingolipid (GSL)-binding motif within the N2 region that may regulate membrane association and ligand endocytosis in *Drosophila* (Hamel et al. 2010). In addition a conserved motif called the Delta and OSM-11-like proteins (DOS) domain has been identified as part of the first two EGF repeats proceeding the DSL domain and is thought to contribute to Notch receptor binding and signaling (Cordle et al. 2008; Komatsu et al. 2008; Parks et al. 2006; Shimizu et al. 1999). Interestingly Dll3 and Dll4 both lack this DOS domain, and therefore it has been proposed that proper activation of the Notch pathway by these ligands would require an additional DOS domain-containing partner (D'Souza et al. 2010; Komatsu et al. 2008). The highly conserved histidine-268 in the second EGF repeat in Jag1 has also been shown to be important for Notch binding and activation of Notch signaling (Hansson et al. 2010). Following the transmembrane domain, the intracellular domain of Notch ligands is the most dissimilar portion of the whole protein, between the different homologues (Pintar et al. 2007). In this region, they contain multiple lysine residues, which are important for ligand endocytosis and a PSD-95/Dlg/ZO-1-ligand (PDZL) motif that is important for Notch-independent signaling and has been predicted for Dll1, Dll4, and Jag1 (Pintar et al. 2007). Mutations of the lysine residues or deletions of the intracellular domain reduce Notch signaling activity in flies and mammalian cells (Glittenberg et al. 2006; Nichols et al. 2007; Parks et al. 2006). Recently, it has been shown that mutation of lysine-613 in the intracellular domain of Dll1 is important for enabling interactions with Notch. Surprisingly this one mutation in the intracellular domain caused a reduction in multiubiquitinylation of Dll1, reduced Notch1 binding and decreased Notch target gene activation in a transcellular assay (Zhang et al. 2011). It would be interesting to see if the other Notch ligands also have this conserved lysine residue in their intracellular domain, and if it too is important for allowing efficient Notch receptor binding.

Dll3 is the most divergent of all the mammalian Notch ligands it does not have any intracellular lysine residues with which to mediate activation of the Notch receptor (Geffers et al. 2007). Furthermore, Dll3 does not bind Notch (Ladi et al. 2005) and is unable to substitute for the loss of Dll1 in mice (Geffers et al. 2007) suggesting that Dll1 and Dll3 are not functionally compatible.

Similar to Notch receptors, Notch ligands can also undergo proteolytic cleavage, both by ADAM metalloproteases and γ-secretase activity, producing an extracellular cleavage product that could modulate Notch signaling and an intracellular cleavage product that is thought to be involved in inhibiting Notch signaling (D'Souza et al. 2010; Jung et al. 2011).

In addition to the canonical ligands described above, several studies have identified non-canonical ligands that are able to activate the Notch signaling pathway. These proteins do not contain the DSL domain, but are glycosylphosphatidylinositol (GPI)-linked or secreted proteins. In addition, some contain the recently identified DOS domain, and it has been suggested that they cooperate or compete with DSL containing ligands to bind Notch (D'Souza et al. 2010; Komatsu et al. 2008).

5 Notch Ligand Endocytosis

Endocytosis of the Notch ligands, has been suggested to be critical to activate Notch signaling; however, the mechanism of how ligand engagement leads to proteolysis of the Notch protein is not well understood. Several molecules involved in ligand endocytosis have been discovered in *Drosophila*, *Xenopus* and Zebrafish.

5.1 Neuralized and Mindbomb

Neuralized (Neur) was first identified in *Drosophila*, and its loss caused a neurogenic phenotype similar to that seen in Notch and Delta mutants (Lehmann 1983). It was originally thought that Neur-enhanced Notch signaling by relieving *cis*-inhibition by Delta (Deblandre et al. 2001); however, expression of Neur is enhanced in signal sending cells, by Neur-induced endocytosis of the Notch ligand and thus stimulating ligand signaling activity (Le Borgne and Schweisguth 2003b). Neur contains a C-terminal 'really interesting new gene' (RING) finger domain, important for its ubiquitin ligase activity, and two Neuralized homology repeats (NHR)-1 and NHR2 (Deblandre et al. 2001; Lai et al. 2001; Pavlopoulos et al. 2001; Yeh et al. 2001). The NHR1 domain is required for binding and trafficking of Delta from the cytoplasm to the plasma membrane (Commisso and Boulianne 2007). Neur also binds phosphoinositol phosphates (PIPs) though a polybasic region in its N terminus, and is required for Neur-mediated Delta endocytosis (Skwarek et al. 2007). The activity of Neur can be further modified by the action of Bearded proteins: Bearded and Twin of m4 (Tom), are negative regulators of Neur function (Bardin and Schweisguth 2006; He et al. 2009). In mammals there are two homologues of Neuralized: Neuralized1 and Neuralized2, both have been found to interact with and ubiquitinylate Delta (Ruan et al. 2001; Song et al. 2006; Vollrath et al. 2001). Despite the strong neurogenic phenotypes in *Drosophila*, Neur1$-/-$ mice do not display any obvious defects in Notch signaling (Ruan et al. 2001; Vollrath et al. 2001). These data suggested that there might be more E3 ubiquitin ligases that could modify DSL ligands in their intracellular domains.

To this end, studies in Zebrafish identified, Mindbomb1 (Mib1) (Chen and Casey Corliss 2004; Itoh et al. 2003). Similar to Neur, Mib binds, and ubiquitinylates Delta, and in the absence of Mib, mutant cells exhibit an accumulation of Delta on their cell surface (Itoh et al. 2003). There are two Zebrafish homologues: ZMib1 and ZMib2. ZMib1 contains three RING finger domains, where its ubiquitin ligase activity resides, and ANK repeats, which are important for internalization (Chen and Casey Corliss 2004). ZMib2 has two RING finger domains and is able to compensate for the absence of ZMib1 (Zhang et al. 2007). There are two homologues in mammals: Mindbomb1 and Mindbomb2 (Koo et al. 2005a, b). Mib1 has been shown to interact with all the Notch ligands (Koo et al. 2005a), whereas Mib2 has been shown to interact with Jag2 only (Takeuchi et al. 2005).

Although several homologues of Neur and Mib have been isolated from several species, and both show a functional similarity in *Drosophila*, Neur and Mib may have evolved differently in vertebrates. *Drosophila* has one Neur gene (with two splice variants (Commisso and Boulianne 2007)), and two Mib1 homologues, both of which are able to bind and ubiquitinylate Delta and Serrate, thus stimulating ligand endocytosis and signaling activities and for the most part these two proteins are also functionally redundant as seen by gene rescue experiments (Lai et al. 2001, 2005; Pitsouli and Delidakis 2005). In contrast, mammalian Neur1 and Neur2 are dispensable for normal development (Koo et al. 2007; Ruan et al. 2001;

Vollrath et al. 2001), as is Mib2 (Koo et al. 2007), whereas deletion of Mib1 produces an embryonic lethal phenotype, and defects in neural tube and brain development and reduced Notch target gene expression (Barsi et al. 2005; Koo et al. 2005a, 2007). Mib2 is not expressed during embryonic development, and thus may explain its inability to compensate for the absence of Mib1 (Koo et al. 2007). Recent studies in *Drosophila* have identified three distinct motifs in the intracellular domain of Delta that are important for Mib1 and Neur docking. Physical interaction with Delta is a prerequisite for its ubiquitinylation, lysine-742 is important for Neur-mediated ubiquitinylation, whereas Mib1 does not seem to prefer any specific lysine residue, this of course may lead to differential trafficking behavior (Daskalaki et al. 2011). The DSL-Mib1 motifs were found to be conserved in other species, and therefore may play an important role in vertebrates too, whereas there was less conservation for the Neur motifs, which supports the lack of functional conservation of Neur in vertebrates (Daskalaki et al. 2011).

Previous reports have also found conserved Mib1 and Neur binding motifs in Serrate (Fontana and Posakony 2009; Glittenberg et al. 2006). Thus, Neur and Mib proteins appear to have similar roles in *Drosophila*, but not necessarily functionally equivalent roles in mammals. Neur2 and Mib1 display different subcellular localization and therefore may have different, yet complementary roles in mammals (Song et al. 2006). It is thought that Mib plays a more direct role in ligand endocytosis in mammals, whereas Neur has been suggested to direct internalized proteins for degradation (Song et al. 2006). It has not been identified whether these differences are due to differential ubiquitinylation states of the Notch ligand intracellular domains. Notch ligands contain several lysine residues (Pintar et al. 2007) that could be mono-, multi-, or polyubiquitinylated. While polyubiqutinylation leads to proteosomal degradation of a protein, mono-, or multiubiquitinylation, leads to endocytosis and intracellular trafficking of the protein (Staub and Rotin 2006). Recently Zhang et al., showed that a specific lysine residue (K613) in the cytoplasmic tail of Dll1 is important for multi-ubiquitinylation, when this lysine was mutated, multi-ubiquitinylation was attenuated and Dll1 only underwent mono-ubiquitinylation, but surprisingly this was not due to the inability of the mutant to interact with Mib1 (Zhang et al. 2011).

5.2 Other Molecules Involved in Endocytosis

Prior to the discovery of Mib and Neur proteins, there was genetic evidence that endocytosis played an important role during Notch signaling. Dynamin (encoded by *Shibre*) is required in both the signal-receiving as well as the signal-sending cells in *Drosophila*. The GTPase Dynamin is important for 'pinching off' endocytic vesicles from the plasma membrane (Seugnet et al. 1997). In addition to Dynamin, other molecules found to be involved in ligand endocytosis are: Auxilin, Rab11, and Clathrin (Banks et al. 2011; Eun et al. 2008; Kandachar et al. 2008). Epsin also seems to play an important role in Notch ligand endocytosis.

Liquid facets (*Lqf*), which encodes *Drosophila* Epsin, binds to Clathrin and other ubiquitinylated cargo (Overstreet et al. 2004). There are three Epsin homologues in mammals, two of which, Epsin 1 and 2 have been deleted in mice, exhibiting an embryonic lethal phenotype and Notch signaling-related defects (Chen et al. 2009).

5.3 Models Proposed for Ligand Endocytosis

There are two main models that have been proposed to explain the purpose of Notch ligand endocytosis and its significance and requirement in inducing Notch receptor activation.

5.3.1 The Mechanotransduction Model

The 'Mechanotransduction' or 'Pulling' model, was first proposed after the observation of the Notch extracellular domain (NECD) in vesicular structures in *Drosophila* signal-sending cells (Nichols et al. 2007; Parks et al. 2000). Losses in Notch ligand endocytosis result in the absence of NECD in the signal-sending cell (Nichols et al. 2007; Parks et al. 2000). The fact that endocytosis is necessary for the activation of the Notch signaling pathway and the above observation, led Parks et al., to suggest that the interaction of the Notch receptor with DSL Notch ligands, was to induce a conformational change in the non-covalent interactions of the heterodimeric Notch receptor that leads to proteolysis by ADAM metalloproteases (Parks et al. 2000). This would release the extracellular portion of the Notch receptor, thus allowing it to be taken up by the signal-sending cell, leaving the transmembrane associated intracellular portion of Notch to undergo γ-secretase mediated cleavage. This process of NECD uptake by the signal-sending cell, is termed 'trans-endocytosis' (Klueg et al. 1998). These observations were confirmed in mammalian cells for Dll1 and Jag1 (Nichols et al. 2007; Parks et al. 2000), blocking of the S2 cleavage did not affect Notch trans-endocytosis, suggesting that Notch ligand endocytosis must create a physical force in order to expose the site for ADAM-mediated cleavage (Hansson et al. 2010; Nichols et al. 2007). Furthermore, structural studies using X-ray crystallography, have shown that the S2 cleavage site is buried deep within the heterodimerization domain and is protected by three LNR domains, suggesting that Notch ligand endocytosis is definitely required to 'stretch' the Notch protein and expose the cleavage site for proteolysis (Gordon et al. 2007, 2008, 2009). In addition, atomic force microscopy has shown that Notch and DSL ligands have a strong binding force (Ahimou et al. 2004). This force would presumably have to be created in the signal-sending cell and involve other proteins, such as Epsin, Dynamin, and Clathrin, which have all been implicated in generating a mechanical force (Liu et al. 2010; McMahon and Gallop 2005) and shown to have a role in Notch ligand endocytosis. Notch receptor binding to ligand induces ubiquitinylation of the Notch ligand, and recruitment of Epsin (Overstreet et al. 2004; Tian et al. 2004; Wang and Struhl 2004). Ligand clustering would presumably increase this assembly of multiple Epsin molecules, and generate a mechanical force to induce endocytosis. By creating a complex that induces Notch ligand endocytosis, and therefore NECD trans-endocytosis, Notch signaling is activated in the signal-receiving cell.

To support this theory proteolysis of Notch ligands causes the release of possible signaling fragments into the extracellular milieu (Mishra-Gorur et al. 2002); however, most secreted forms of DSL ligands are unable to induce Notch signaling, presumably due to the absence of a pulling force not being generated. The ability of Notch ligands to signal when immobilized, cross-linked, or clustered, suggests that a force is being generated to induce Notch activation (Hicks et al. 2002; Varnum-Finney et al. 2000).

5.3.2 The Recycling Model

The Recycling model is based on the observation that bulk endocytosis of Delta is unaffected in *Drosophila* Epsin mutants (*Lqf*−), yet Delta is unable to signal, the *Lqf*− mutant exhibits phenotypes similar to the loss of Notch, Delta, and Serrate, and therefore is thought to be required by the signal-sending cell (Wang and Struhl 2004). Wang and Struhl suggested that the newly synthesized Delta was not able to signal as it had not been modified, and that it required internalization and sorting, a process that required ubiquitinylation and then trafficking back to the plasma membrane, after having been post-translationally modified or relocalized to a more favorable location for signaling. This model suggests that internalization and recycling cause the ligand to become 'activated' and having acquired its signaling capabilities, is better able to interact with the Notch receptor (Wang and Struhl 2004). Removal of the intracellular domain of Delta inhibits endocytosis; however, replacing this domain either with the intracellular domain of the low density lipoprotein (LDL) receptor, which is known to be internalized by endocytosis, or an extendable ubiquitin (Ub), which can be polyubiquitinylated, promotes internalization and signaling activity in *Drosophila* (Wang and Struhl 2004), whereas a non-extendable Ub signals more weakly (Wang and Struhl 2004).

In mammalian cells, mutation of all 17 lysine residues in Dll1 (Dll1K17), reduces the ability of this construct to internalize, and transendocytose the Notch receptor, in addition it renders this mutant unable to recycle back to the cell surface, (Heuss et al. 2008). These authors suggest that the process involves localization to the lipid rafts and that recycling increases the affinity of Dll1 for Notch binding (Heuss et al. 2008). In contrast, Zhang et al., show that mutation of a single lysine, K613, in the intracellular domain of Dll1 enhances lipid raft localization (Zhang et al. 2011), suggesting that different lysine residues may be important for different functions with regards to ligand localization and endocytosis.

Initially several Notch loss-of-function phenotypes and Delta recycling defects were identified that seem to support this notion: Sec15, a component of the exocyst complex in *Drosophila*, and a dominant-negative form of Rab11 in mammalian cells, lead to recycling defects (Emery et al. 2005; Jafar-Nejad et al. 2005); however, recently it has been shown that Sec15, which functions with Rab11, is not required in every Notch signaling event (Banks et al. 2011; Eun et al. 2008; Jafar-Nejad et al. 2005; Windler and Bilder 2010).

If the recycling model is the method of endocytosis that activates the Notch signaling pathway, then further questions as to what leads to an 'activated' Notch ligand need to be identified, suggestions include: clustering of ligands, trafficking into lipid microdomains, proteolytic cleavages, and post-translational modifications (Chitnis 2006; Le Borgne and Schweisguth 2003a; Wang and Struhl 2004).

The Mechanotransduction model and Recycling models are not necessarily mutually exclusive. It has been suggested that perhaps two ligand internalization events are required, the first to activate the ligand, and the second to activate the receptor and induce Notch signaling in the signal-receiving cell. Heuss et al., showed that mutation of all 17 lysine residues in Dll1 (Dll1K17), reduces the ability of this construct to internalize, and recycle back to the cell surface (Heuss et al. 2008). These authors suggest that the process involves localization to the lipid rafts and that recycling increases the affinity of Dll1 for Notch binding (Heuss et al. 2008); however, the same group also showed that this may not be entirely due to the lack of ubiquitinylation, as a lysine-less intracellular domain of Dll3, when fused to the extracellular domain of Dll1 (Dll1-3chimera), is still able to undergo internalization and recycling, but is unable to activate Notch signaling, despite its ability to bind Notch, suggesting other mechanisms are important in Notch ligands signaling capabilities (Heuss et al. 2008). Studies using a mutant form of Jag1, also lend support to both models. A mutant form of Jag1, Jag^{Ndr}, which has a missense mutation in the extracellular domain at position histidine-268 located in the second EGF repeat, has lost the ability to bind and activate Notch, providing support for the Mechanotransduction model. In this report, Jag1 and NECD were targeted for degradation, whereas Jag^{Ndr} expressing cells fail to trans-endocytose NECD. In the absence of Notch, trafficking and interaction with Mib1 were similar to WT Jag1. siRNA-mediated inhibition of MIB1 function led to a loss of receptor activation, but not receptor binding by WT Jag1, providing support for the Recycling model (Hansson et al. 2010).

The role of Jag2 and Dll4 in activating the Notch signaling pathway via its endocytosis has yet to be fully determined, although deletion of the intracellular domain of Dll4 is unable to activate Notch signaling in a functional assay (Abe et al. 2010). Further analysis through deletions and mutations of Notch ligands will help to answer this question.

6 Modulators of Notch Signaling

Notch signaling is regulated at multiple levels in the pathway. Notch receptors can be modified by the action of Fringe glycosyl transferases, which can influence ligand specificity (Stanley and Okajima 2010). In the cytoplasm, proteins such as

Numb interact with Notch and promote its ubiquitinylation and degradation (Guo et al. 1996; Rhyu et al. 1994; Spana and Doe 1996; Uemura et al. 1989; Zhong et al. 1997). The mechanism underlying this inhibition involves the E3 ubiquitin ligase, Itch (McGill and McGlade 2003), which also interacts with and ubiquitinylates Notch (Fang and Kerppola 2004). In the nucleus, Sel-10/Fbw7 is an E3 ubiquitin ligase and member of the F box family of proteins, that interacts with a cyclin dependent kinase 8 (CDK8) phosphorylated nuclear form of Notch, and targets it for ubiquitinylation and degradation by the proteosome (Fryer et al. 2004; Gupta-Rossi et al. 2001; Oberg et al. 2001). Deltex (Matsuno et al. 2002), Nrarp (Notch-regualted ankyrin repeat protein) (Lamar et al. 2001), and MINT (also known as SHARP) (Kuroda et al. 2003) are also involved in Notch regulation, and together may serve to fine-tune the activity of Notch within the signal-receiving cell.

7 Notch Signaling and T Cell Development

7.1 Notch Receptor Expression in the Thymus

Notch-1, -2, and -3 receptor proteins are all expressed in thymocytes (Felli et al. 1999; Fiorini et al. 2009; Hasserjian et al. 1996; Shi et al. 2011). Notch1 expression is differentially expressed throughout T cell development. The use of novel monoclonal antibodies, against Notch-1 and Notch-2 confirmed that Notch1 expression is highest in DN and immature single positive (ISP) $CD8^+$ thymocyte subsets in adult and embryonic stage 16 (E16), and decreases steadily in DP cells in adults (Fiorini et al. 2009), with intermediate to low expression in SP cells (Fiorini et al. 2009; Hasserjian et al. 1996). Notch2 expression was found to be highest in DN1 and DN2 subsets of adult thymocyte preparations (Fiorini et al. 2009). In addition, Notch1 and -2 are expressed in $\gamma\delta$ T cells isolated from adult mice (Fiorini et al. 2009). Notch3 mRNA and protein expression is found in DN and DP thymocyte subsets, with the highest levels in DN3 cells (Shi et al. 2011).

It is well established that Notch signaling is critically required for the T vs. B cell lineage decision (Han et al. 2002; Izon et al. 2002; Maillard et al. 2004; Pui et al. 1999; Radtke et al. 1999). Continuous Notch signals are required for the commitment of the earliest thymic progenitor cells to the DN3 stage of T cell development (Schmitt et al. 2004), VDJ recombination (Wolfer et al. 2002) and the β-selection checkpoint (Ciofani et al. 2004; Maillard et al. 2006; Tanigaki et al. 2004), for its functional outcomes, including: survival, proliferation, and differentiation (Ciofani et al. 2004; Ciofani and Zuniga-Pflucker 2005). Notch signaling is also required at the $\alpha\beta$ and $\gamma\delta$ lineage decision (Ciofani et al. 2006; Robey et al. 1996; Washburn et al. 1997; Wolfer et al. 2001, 2002). Its role in the CD4 vs. CD8 lineage decision remains controversial (Robey et al. 1996; Wolfer et al. 2001); however, the Notch and TCR signaling pathways may work together to influence positive selection and CD4/CD8 T cell development as evidenced in mice conditionally deleted for Presenilin1/2 (Laky and Fowlkes 2007).

Although much is known about the role of Notch1 in T cell development, recent studies have focused on the role of Notch3. Recent data has found that Notch3 is dispensable for T cell development (Shi et al. 2011; Suliman et al. 2011), although previous studies had suggested a role for Notch3 in T cell development (Bellavia et al. 2000). Notch3 was found to have a lower binding affinity for Dll4 (Suliman et al. 2011) and therefore presumably does not interact with Dll4 in the thymus. However, Petrie and colleagues examined the same mice, and found that Notch3 gene trapped mice had slightly smaller thymi, although this was not statistically significant. They reasoned that the mild phenotype observed may be due to a compensatory role by Notch1, however, this was not the case, as deletion of Notch3 did not exacerbate the Notch1 phenotype (Shi et al. 2011).

7.2 Notch Ligand Expression in the Thymus

In vivo, analysis of Notch ligand expression has revealed that Dll1, Dll4, Jag1, and Jag2 are expressed in the day 16 embryonic thymus, with Dll4 exhibiting the highest level of expression (Mohtashami et al. 2010). In adults, Dll4 is most highly expressed in cortical thymic epithelial cells (cTECs) (Hozumi et al. 2008; Koch et al. 2008), in particular in the subcapsular and outer cortical areas of the thymus (Mohtashami et al. 2010), which are incidentally where DN subsets express high levels of Notch1 and are undergoing β-selection. Dll1 expression was found to be absent from inner cortical or medullary regions, and confined to the blood vessels, the point of entry of blood-borne progenitors; (Koch et al. 2008) however, other reports have found Dll1 expressed in the cortex and pericortex (Mohtashami et al. 2010; Schmitt et al. 2004). Recently, Dll3 expression was observed in DN, $CD4^+$, and $CD8^+$ thymocytes, with the highest expression in DN3 subsets (Hoyne et al. 2011). In adults, Jag1 is expressed at low levels throughout the thymus, whereas Jag2 is expressed most highly in the inner cortex and outer medulla (Hozumi et al. 2008; Koch et al. 2008).

Despite the expression of several Notch ligands in the thymus, it is now clear that Dll4 is the relevant physiological ligand within the thymus (Hozumi et al. 2008; Koch et al. 2008). Conditional deletion of Dll4 abrogated T cell development and led to the development of B cells in the thymus (Hozumi et al. 2008; Koch et al. 2008), in contrast, conditional deletion of Dll1 did not inhibit T cell development, and presumably Dll4 was able to compensate for its absence (Hozumi et al. 2004). Jag1−/− mice and Jag2 mutant mice, exhibit normal $\alpha\beta$ T cell development (Jiang et al. 1998; Mancini et al. 2005), however, Jag2 mutant mice have reduced numbers of $\gamma\delta$ thymocytes (Jiang et al. 1998). Interestingly Hoyne et al. have recently reported a cell autonomous role for Dll3 in the thymus, with Dll3−/− mice exhibiting increased *hes5* transcription at the DP stage, thus suggesting an inhibitory role in Notch signaling (Hoyne et al. 2011).

The transcription factor (TF) FoxN1, which is required for TEC differentiation (Su et al. 2003), also seems to regulate the expression of Dll1, Dll4, and Jag2 expression in TECs. Mice expressing a hypomorphic allele of FoxN1, exhibited decreased mRNA expression of these Notch ligands (Xiao et al. 2008).

Although Dll4 is the physiological ligand, in vitro studies have shown that T cell development can be supported from the other Notch ligands too, such as with the OP9-DL1 system (Zuniga-Pflucker 2004), the results of which are presumably precluded in vivo, by the presence of Dll4 (Hozumi et al. 2004). Several labs have shown that T cell development can be supported from over expression of Dll1 in vitro (Besseyrias et al. 2007; Jaleco et al. 2001; Schmitt and Zuniga-Pflucker 2002). In addition, experiments carried out by Radtke and colleagues showed that analysis of Notchl conditionally deleted BM derived precursors, which are unable to give rise to T cells in vivo, were able to give rise to T cells in vitro when cultured on OP9-DL1 stromal layers, indicating that this was a Notch-2-mediated event. While, Notch1/Notch2 conditionally deleted BM-derived precursors were not able to give rise to T cells when grown on OP9-DL1 or OP9-DL4, indicating that Notch2 is only able to interact with Dll1, whereas, Notch1 is able to interact with both Dll1 and Dll4 (Besseyrias et al. 2007).

To assess whether Dll1 and Dll4 are equally capable at supporting T cell development, Mohtashami et al., recently showed that Dll4 is better able to support T cell development in vitro compared to Dll1. These authors directly compared the ability of these two proteins to induce and support T cell development, when expressed in OP9 cells (Mohtashami et al. 2010). Similar to previous reports, when Dll1 and Dll4 are expressed at high levels in OP9 cells, both proteins are able to induce and support $\alpha\beta$ and $\gamma\delta$ T cell development (Besseyrias et al. 2007; Schmitt and Zuniga-Pflucker 2002). However, when Dll4 is expressed at lower levels in OP9 cells, but similar to the levels on thymic epithelial cells, Dll4 is better able to support T and myeloid cell development and induce Notch- and myeloid-specific gene transcription, from fetal liver (FL)-derived progenitors, compared to Dll1, which exhibited an increase in the development of myeloid and B cell lineages (Mohtashami et al. 2010).

Overexpression of Jag1 in stromal cells was unable to give rise to $\alpha\beta$ T cells, but was able to support $\gamma\delta$ T cell and NK cell lineages (Lehar et al. 2005), suggesting a role for Jagged proteins in supporting $\gamma\delta$ T cell development. Recent data from Habu's group confirms that Jag1 is unable to support T cell development, but they did show that Jag2 is able to support T cell development from FL-derived HSCs, in vitro, to the DP stage, when cocultured on transfected PA6 cell lines, and that this was dependent on the magnitude of the signal available (Abe et al. 2010).

These various in vivo and in vitro studies highlight differences in Notch ligands and their ability to support T cell development, and these differences may be related to their structure and their ability to induce Notch receptor activation. One of the proposed methods of Notch activation lies in the ability of the endocytosis of these Notch ligands, suggesting that this could explain some of the differences described above. Notch ligands have the least similarity in their intracellular domain (Pintar et al. 2007). The intracellular domain is the target for Mib and

Neur-mediated ubiquitinylation, inducing endocytosis of the Notch ligands and activation of the Notch signaling pathway (D'Souza et al. 2010; Yamamoto et al. 2010). Recent studies focusing on endocytosis of Notch ligands and other molecules involved in this process have shed some light on this area, with regards to T cell development.

7.3 Notch Ligand Endocytosis and Role in T Cell Development

In vitro studies have primarily focused on the role of Dll1 endocytosis in T cell development. Heuss et al., showed that when Dll1, that has been mutated at all 17 potential lysine residues in its intracellular domain (Dll1K17), is overexpressed in OP9 cells, it is unable to support T cell development from FL-derived progenitors to the DP stage, producing 12-fold lower DP cells than OP9 cells expressing wild-type Dll1 (Heuss et al. 2008). These results were also observed with a Dll1-Dll3 chimeric construct, which is unable to be ubiquitinylated. Both constructs were unable to trans-endocytose NECD efficiently (Heuss et al. 2008), indicating that ubiquitinylation is a requirement for ligand endocytosis and thus Notch activation in developing progenitors to differentiate and adopt the T cell lineage. Further studies that investigate the role of the intracellular domain of Notch ligands come from Habu's group. Here, Habu and colleagues showed that the Dll4 and Jag1 intracellular domains are necessary, but interchangeable and that the signaling capabilities of these two Notch ligands lie in the interactions of their extracellular domains with Fringe-modified Notch receptors (Abe et al. 2010). FL HSCs cocultured on transfected PA6 cells with Jag1 extracellular domain fused to Dll4 intracellular domain (J1d-D4), were unable to support T cell development, whereas when the Dll4 extracellular domain fused with the Jag1 intracellular domain is overexpressed in either NIH-3T3 or PA6-derived transfectants, T-lineage cells are able to develop in these cultures (Abe et al. 2010). These authors also reported that overexpression of a Dll4 construct that lacks its intracellular domain, is unable to support T cell development (Abe et al. 2010). Swapping of the DSL domain of Dll4 or Jag2 onto Jag1 did not confer their T cell inducing capabilities onto Jag1, and suggested that additional sequences within the Jag2 protein were needed for effective signaling to promote T cell development (Abe et al. 2010).

Studies focused on the E3 ubiquitin ligases which ubiquitinylate the intracellular domain of the Notch ligands (Daskalaki et al. 2011; Itoh et al. 2003), show that *mib1*, *mib2*, *neur1*, and *neur2* are all expressed in both the thymic stromal and $CD45^+$ hematopoietic compartments (Song et al. 2008), with Mib1 and Mib2 exhibiting the highest level of expression. Conditional deletion of Mib1 by Cre under the control of the MMTV promoter caused a reduction in Mib1 expression in the thymic stromal cells. Deletion of Neur1, Neur2, or Mib2 did not affect T cell development (Song et al. 2008); however, conditional deletion of Mib1 by Cre under the MMTV promoter or Mx1, interferon inducible promoter, led to the abrogation of T cell development in these mice (Song et al. 2008), indicating that Mib1 is the relevant physiological E3 ubiquitin ligase in the thymus. These mice exhibited a reduction in total cellularity, decreased ETP, DP, and SP cell populations, and an increase in B220^{+} B-lineage cells in the DN compartment of these mice. To confirm that Mib1 acts in a non-cell autonomous manner, WT BM cells were unable to reconstitute the Mib1−/− conditionally deleted thymus, whereas Mib1 deficient BM cells were able to reconstitute a WT thymus, showing that Mib1's activity resides in the thymic epithelial cells (Song et al. 2008). Dll1 endocytosis was normal in Neur1−/−, Neur2−/−, and Mib2−/− thymi, whereas Dll1 accumulated on cortical thymic epithelial cells in the Mib1 conditionally deleted thymus, indicating its importance. Furthermore, siRNA knockdown of Mib1 in OP9 stromal cells that overexpressed Dll1 caused a block in development of BM LSKs or FL LSKs at the DN1 stage, indicating that Mib1 can interact with Dll1 (Song et al. 2008).

The ubiquitinylation of Dll ligands is a requirement for endocytosis and currently there is little data for the ubiquitinylation of Dll4 and its role in T cell development. As Dll4 is the physiological ligand, it would give an insight into its regulation. In particular, in light of Mib1 being the functional E3 ubiquitin ligase in the thymus, it would be interesting to investigate the role of Mib1 on Dll4, presumably Mib1 interacts with and ubiquitinylates Dll4 in vivo.

7.4 Notch Modulators and T Cell Development

Mice conditionally deleted for Lunatic Fringe (Lfng) under the Lck promoter, exhibit a decrease in the transition from DN3 to DP cell development (Yuan et al. 2011), whereas overexpression of Lfng in the thymus led to a decrease in the generation of T cells in the thymus, and the development of intrathymic B cells (Koch et al. 2001). The same group showed that DN3 and DN4 cells were found to exhibit increased proliferation and self-renewal when Lfng was overexpressed, thereby increasing the amount of time that β-selected pre-DP cells were exposed to OP9-DL induced proliferation (Yuan et al. 2011). Overall these studies show that Lfng-mediated glycosylation directly increases Notch1 binding affinity for Dll1 and Dll4, and recently also Jag2, but decreases their affinity for Jag1 (Van de Walle et al. 2011).

Inactivation of Numb or Numb-like in T cells indicates that they are dispensable for T cell development (Anderson et al. 2005; Wilson et al. 2007). This is in contrast to other studies, in which overexpression of Numb specifically in T cells causes a reduction in Notch activity (French et al. 2002); however, recent studies using overexpression of full length Numb resulted in a larger thymus and loss of pre-TCR signaling, whereas overexpression of a dominant-negative Numb resulted in a smaller thymus and enhanced pre-TCR signaling and differentiation (Aguado et al. 2010) indicating that the role of Numb in T cell development still requires further analysis.

ADAM proteases are required for both Notch receptor and ligand proteolytic processing. Several ADAM proteases are expressed in the thymus and some have demonstrated a role in T cell development (Haidl et al. 2002; Li et al. 2007). Recently data from several labs show that the absence of ADAM8 causes an accumulation of mature thymocytes and increased medulla size, due to increased proliferation, decreased apoptosis and impairment in thymocyte migration (Gossens et al. 2010). Mice conditionally deleted for ADAM10 by Cre under the control of the Lck promoter showed a reduction in total thymic output, a reduction in DP and SP thymocytes and decrease in the Notch target gene *deltex1* (Tian et al. 2008). Furthermore, overexpression of ADAM10 in early lymphocyte progenitors under the H-2 Kb promoter and the IgH enhancer region, exhibited a smaller thymus, reduced DN and DP thymocyte populations and other B and myeloid specific defects (Gibb et al. 2011). In contrast, proteolytic processing of Notch ligands is not critical during T cell development. Using an uncleavable Dll1 overexpressed in OP9 cells, T cell development was analyzed and found to progress normally (Gravano and Manilay 2010).

8 Summary

A complex interplay of various factors is required to initiate and support the development of T cells. Inherent to this complexity is the requirement for Notch signaling. Recent studies have highlighted the importance of Notch ligand endocytosis in activating the Notch signal, and importantly have identified some of the key physiological players within the thymus. The task now is to determine the precise interactions that some of these proteins have in the thymus, in particular in the regulation and interaction between the Notch ligand Dll4 and the E3 ubiquitin ligase Mib1. Several questions remain: Does Dll4 require endocytosis for Notch activation during T cell development? Is it regulated in the same manner as the other Notch ligands? Which lysine residues serve as potential ubiquitinylation sites on Dll1 and Dll4 by Mib1? Is a lysine-613-like residue also important for Dll4? Does it also affect Notch binding of Dll4? What other DOS-containing ligands are expressed in the thymus? Are any of these important in potentiating the Notch signal by canonical DSL Notch ligands? Which RING finger domain in Mib1 is important for interacting with the Notch ligands in the thymus? Further studies will certainly begin to address these questions to better understand the role of Notch ligand endocytosis in the thymus that predicates the induction and support of T-lineage commitment and differentiation.

References

Abe N, Hozumi K, Hirano K, Yagita H, Habu S (2010) Notch ligands transduce different magnitudes of signaling critical for determination of T-cell fate. Eur J Immunol 40:2608–2617

Acar M, Jafar-Nejad H, Takeuchi H, Rajan A, Ibrani D, Rana NA, Pan H, Haltiwanger RS, Bellen HJ (2008) Rumi is a CAP10 domain glycosyltransferase that modifies Notch and is required for Notch signaling. Cell 132:247–258

Aguado R, Martin-Blanco N, Caraballo M, Canelles M (2010) The endocytic adaptor Numb regulates thymus size by modulating pre-TCR signaling during asymmetric division. Blood 116:1705–1714

Ahimou F, Mok LP, Bardot B, Wesley C (2004) The adhesion force of Notch with Delta and the rate of Notch signaling. J Cell Biol 167:1217–1229

Anderson AC, Kitchens EA, Chan SW, St Hill C, Jan YN, Zhong W, Robey EA (2005) The Notch regulator Numb links the Notch and TCR signaling pathways. J Immunol 174:890–897

Artavanis-Tsakonas S, Muskavitch MA, Yedvobnick B (1983) Molecular cloning of Notch, a locus affecting neurogenesis in *Drosophila melanogaster*. Proc Natl Acad Sci U S A 80:1977–1981

Banks SM, Cho B, Eun SH, Lee JH, Windler SL, Xie X, Bilder D, Fischer JA (2011) The functions of auxilin and Rab11 in *Drosophila* suggest that the fundamental role of ligand endocytosis in notch signaling cells is not recycling. PLoS ONE 6:e18259

Bardin AJ, Schweisguth F (2006) Bearded family members inhibit Neuralized-mediated endocytosis and signaling activity of Delta in *Drosophila*. Dev Cell 10:245–255

Barsi JC, Rajendra R, Wu JI, Artzt K (2005) Mind bomb1 is a ubiquitin ligase essential for mouse embryonic development and Notch signaling. Mech Dev 122:1106–1117

Bellavia D, Campese AF, Alesse E, Vacca A, Felli MP, Balestri A, Stoppacciaro A, Tiveron C, Tatangelo L, Giovarelli M, Gaetano C, Ruco L, Hoffman ES, Hayday AC, Lendahl U, Frati L, Gulino A, Screpanti I (2000) Constitutive activation of NF-kappaB and T-cell leukemia/ lymphoma in Notch3 transgenic mice. EMBO J 19:3337–3348

Besseyrias V, Fiorini E, Strobl LJ, Zimber-Strobl U, Dumortier A, Koch U, Arcangeli ML, Ezine S, Macdonald HR, Radtke F (2007) Hierarchy of Notch-Delta interactions promoting T cell lineage commitment and maturation. J Exp Med 204:331–343

Bettenhausen B, Hrabe de Angelis M, Simon D, Guenet JL, Gossler A (1995) Transient and restricted expression during mouse embryogenesis of Dll1, a murine gene closely related to *Drosophila* Delta. Development 121:2407–2418

Bhandoola A, von Boehmer H, Petrie HT, Zuniga-Pflucker JC (2007) Commitment and developmental potential of extrathymic and intrathymic T cell precursors: plenty to choose from. Immunity 26:678–689

Blaumueller CM, Qi H, Zagouras P, Artavanis-Tsakonas S (1997) Intracellular cleavage of Notch leads to a heterodimeric receptor on the plasma membrane. Cell 90:281–291

Bray SJ (2006) Notch signalling: a simple pathway becomes complex. Nat Rev Mol Cell Biol 7:678–689

Chen W, Casey Corliss D (2004) Three modules of zebrafish Mind bomb work cooperatively to promote Delta ubiquitination and endocytosis. Dev Biol 267:361–373

Chen H, Ko G, Zatti A, Di Giacomo G, Liu L, Raiteri E, Perucco E, Collesi C, Min W, Zeiss C, De Camilli P, Cremona O (2009) Embryonic arrest at midgestation and disruption of Notch signaling produced by the absence of both epsin 1 and epsin 2 in mice. Proc Natl Acad Sci U S A 106:13838–13843

Chitnis A (2006) Why is delta endocytosis required for effective activation of notch? Dev Dyn 235:886–894

Ciofani M, Zuniga-Pflucker JC (2005) Notch promotes survival of pre-T cells at the beta-selection checkpoint by regulating cellular metabolism. Nat Immunol 6:881–888

Ciofani M, Zuniga-Pflucker JC (2007) The thymus as an inductive site for T lymphopoiesis. Annu Rev Cell Dev Biol 23:463–493

Ciofani M, Schmitt TM, Ciofani A, Michie AM, Cuburu N, Aublin A, Maryanski JL, Zuniga-Pflucker JC (2004) Obligatory role for cooperative signaling by pre-TCR and Notch during thymocyte differentiation. J Immunol 172:5230–5239

Ciofani M, Knowles GC, Wiest DL, von Boehmer H, Zuniga-Pflucker JC (2006) Stage-specific and differential notch dependency at the alphabeta and gammadelta T lineage bifurcation. Immunity 25:105–116

Commisso C, Boulianne GL (2007) The NHR1 domain of Neuralized binds Delta and mediates Delta trafficking and Notch signaling. Mol Biol Cell 18:1–13

Cordle J, Johnson S, Tay JZ, Roversi P, Wilkin MB, de Madrid BH, Shimizu H, Jensen S, Whiteman P, Jin B, Redfield C, Baron M, Lea SM, Handford PA (2008) A conserved face of the Jagged/Serrate DSL domain is involved in Notch trans-activation and cis-inhibition. Nat Struct Mol Biol 15:849–857

D'Souza B, Meloty-Kapella L, Weinmaster G (2010) Canonical and non-canonical Notch ligands. Curr Top Dev Biol 92:73–129

Daskalaki A, Shalaby NA, Kux K, Tsoumpekos G, Tsibidis GD, Muskavitch MA, Delidakis C (2011) Distinct intracellular motifs of Delta mediate its ubiquitylation and activation by Mindbomb1 and Neuralized. J Cell Biol 195:1017–1031

Deblandre GA, Lai EC, Kintner C (2001) Xenopus neuralized is a ubiquitin ligase that interacts with XDelta1 and regulates Notch signaling. Dev Cell 1:795–806

Deftos ML, Huang E, Ojala EW, Forbush KA, Bevan MJ (2000) Notch1 signaling promotes the maturation of CD4 and CD8 SP thymocytes. Immunity 13:73–84

Del Amo FF, Smith DE, Swiatek PJ, Gendron-Maguire M, Greenspan RJ, McMahon AP, Gridley T (1992) Expression pattern of Motch, a mouse homolog of *Drosophila* Notch, suggests an important role in early postimplantation mouse development. Development 115:737–744

Dunwoodie SL, Henrique D, Harrison SM, Beddington RS (1997) Mouse Dll3: a novel divergent Delta gene which may complement the function of other Delta homologues during early pattern formation in the mouse embryo. Development 124:3065–3076

Ellisen LW, Bird J, West DC, Soreng AL, Reynolds TC, Smith SD, Sklar J (1991) TAN-1, the human homolog of the *Drosophila* notch gene, is broken by chromosomal translocations in T lymphoblastic neoplasms. Cell 66:649–661

Emery G, Hutterer A, Berdnik D, Mayer B, Wirtz-Peitz F, Gaitan MG, Knoblich JA (2005) Asymmetric Rab 11 endosomes regulate delta recycling and specify cell fate in the *Drosophila* nervous system. Cell 122:763–773

Eun SH, Banks SM, Fischer JA (2008) Auxilin is essential for Delta signaling. Development 135:1089–1095

Fang D, Kerppola TK (2004) Ubiquitin-mediated fluorescence complementation reveals that Jun ubiquitinated by Itch/AIP4 is localized to lysosomes. Proc Natl Acad Sci U S A 101: 14782–14787

Felli MP, Maroder M, Mitsiadis TA, Campese AF, Bellavia D, Vacca A, Mann RS, Frati L, Lendahl U, Gulino A, Screpanti I (1999) Expression pattern of Notch1, 2 and 3 and Jagged1 and 2 in lymphoid and stromal thymus components: distinct ligand-receptor interactions in intrathymic T cell development. Int Immunol 11:1017–1025

Fiorini E, Merck E, Wilson A, Ferrero I, Jiang W, Koch U, Auderset F, Laurenti E, Tacchini Cottier F, Pierres M, Radtke F, Luther SA, Macdonald HR (2009) Dynamic regulation of notch 1 and notch 2 surface expression during T cell development and activation revealed by novel monoclonal antibodies. J Immunol 183:7212–7222

Fontana JR, Posakony JW (2009) Both inhibition and activation of Notch signaling rely on a conserved Neuralized-binding motif in Bearded proteins and the Notch ligand Delta. Dev Biol 333:373–385

French MB, Koch U, Shaye RE, McGill MA, Dho SE, Guidos CJ, McGlade CJ (2002) Transgenic expression of numb inhibits notch signaling in immature thymocytes but does not alter T cell fate specification. J Immunol 168:3173–3180

Fryer CJ, Lamar E, Turbachova I, Kintner C, Jones KA (2002) Mastermind mediates chromatin-specific transcription and turnover of the Notch enhancer complex. Genes Dev 16:1397–1411

Fryer CJ, White JB, Jones KA (2004) Mastermind recruits CycC:CDK8 to phosphorylate the Notch ICD and coordinate activation with turnover. Mol Cell 16:509–520
Geffers I, Serth K, Chapman G, Jaekel R, Schuster-Gossler K, Cordes R, Sparrow DB, Kremmer E, Dunwoodie SL, Klein T, Gossler A (2007) Divergent functions and distinct localization of the Notch ligands DLL1 and DLL3 in vivo. J Cell Biol 178:465–476
Gibb DR, Saleem SJ, Chaimowitz NS, Mathews J, Conrad DH (2011) The emergence of ADAM10 as a regulator of lymphocyte development and autoimmunity. Mol Immunol 48:1319–1327
Glittenberg M, Pitsouli C, Garvey C, Delidakis C, Bray S (2006) Role of conserved intracellular motifs in Serrate signalling, cis-inhibition and endocytosis. EMBO J 25:4697–4706
Gordon WR, Vardar-Ulu D, Histen G, Sanchez-Irizarry C, Aster JC, Blacklow SC (2007) Structural basis for autoinhibition of Notch. Nat Struct Mol Biol 14:295–300
Gordon WR, Arnett KL, Blacklow SC (2008) The molecular logic of Notch signaling—a structural and biochemical perspective. J Cell Sci 121:3109–3119
Gordon WR, Roy M, Vardar-Ulu D, Garfinkel M, Mansour MR, Aster JC, Blacklow SC (2009) Structure of the Notch1-negative regulatory region: implications for normal activation and pathogenic signaling in T-ALL. Blood 113:4381–4390
Gossens K, Naus S, Hollander GA, Ziltener HJ (2010) Deficiency of the metalloproteinase-disintegrin ADAM8 is associated with thymic hyper-cellularity. PLoS ONE 5:e12766
Gravano DM, Manilay JO (2010) Inhibition of proteolysis of Delta-like-1 does not promote or reduce T-cell developmental potential. Immunol Cell Biol 88:746–753
Guo M, Jan LY, Jan YN (1996) Control of daughter cell fates during asymmetric division: interaction of Numb and Notch. Neuron 17:27–41
Gupta-Rossi N, Le Bail O, Gonen H, Brou C, Logeat F, Six E, Ciechanover A, Israel A (2001) Functional interaction between SEL-10, an F-box protein, and the nuclear form of activated Notch1 receptor. J Biol Chem 276:34371–34378
Haidl ID, Huber G, Eichmann K (2002) An ADAM family member with expression in thymic epithelial cells and related tissues. Gene 283:163–170
Hamel S, Fantini J, Schweisguth F (2010) Notch ligand activity is modulated by glycosphingolipid membrane composition in *Drosophila melanogaster*. J Cell Biol 188:581–594
Han H, Tanigaki K, Yamamoto N, Kuroda K, Yoshimoto M, Nakahata T, Ikuta K, Honjo T (2002) Inducible gene knockout of transcription factor recombination signal binding protein-J reveals its essential role in T versus B lineage decision. Int Immunol 14:637–645
Hansson EM, Lanner F, Das D, Mutvei A, Marklund U, Ericson J, Farnebo F, Stumm G, Stenmark H, Andersson ER, Lendahl U (2010) Control of Notch-ligand endocytosis by ligand-receptor interaction. J Cell Sci 123:2931–2942
Hasserjian RP, Aster JC, Davi F, Weinberg DS, Sklar J (1996) Modulated expression of notch1 during thymocyte development. Blood 88:970–976
Hayes SM, Li L, Love PE (2005) TCR signal strength influences alphabeta/gammadelta lineage fate. Immunity 22:583–593
He F, Saito K, Kobayashi N, Harada T, Watanabe S, Kigawa T, Guntert P, Ohara O, Tanaka A, Unzai S, Muto Y, Yokoyama S (2009) Structural and functional characterization of the NHR1 domain of the *Drosophila* neuralized E3 ligase in the notch signaling pathway. J Mol Biol 393:478–495
Heuss SF, Ndiaye-Lobry D, Six EM, Israel A, Logeat F (2008) The intracellular region of Notch ligands Dll1 and Dll3 regulates their trafficking and signaling activity. Proc Natl Acad Sci U S A 105:11212–11217
Hicks C, Ladi E, Lindsell C, Hsieh JJ, Hayward SD, Collazo A, Weinmaster G (2002) A secreted Delta1-Fc fusion protein functions both as an activator and inhibitor of Notch1 signaling. J Neurosci Res 68:655–667
Hoyne GF, Chapman G, Sontani Y, Pursglove SE, Dunwoodie SL (2011) A cell autonomous role for the Notch ligand Delta-like 3 in alphabeta T-cell development. Immunol Cell Biol 89:696–705
Hozumi K, Negishi N, Suzuki D, Abe N, Sotomaru Y, Tamaoki N, Mailhos C, Ish-Horowicz D, Habu S, Owen MJ (2004) Delta-like 1 is necessary for the generation of marginal zone B cells but not T cells in vivo. Nat Immunol 5:638–644

Hozumi K, Mailhos C, Negishi N, Hirano K, Yahata T, Ando K, Zuklys S, Hollander GA, Shima DT, Habu S (2008) Delta-like 4 is indispensable in thymic environment specific for T cell development. J Exp Med 205:2507–2513

Itoh M, Kim CH, Palardy G, Oda T, Jiang YJ, Maust D, Yeo SY, Lorick K, Wright GJ, Ariza-McNaughton L, Weissman AM, Lewis J, Chandrasekharappa SC, Chitnis AB (2003) Mind bomb is a ubiquitin ligase that is essential for efficient activation of Notch signaling by Delta. Dev Cell 4:67–82

Izon DJ, Aster JC, He Y, Weng A, Karnell FG, Patriub V, Xu L, Bakkour S, Rodriguez C, Allman D, Pear WS (2002) Deltex1 redirects lymphoid progenitors to the B cell lineage by antagonizing Notch1. Immunity 16:231–243

Jafar-Nejad H, Andrews HK, Acar M, Bayat V, Wirtz-Peitz F, Mehta SQ, Knoblich JA, Bellen HJ (2005) Sec15, a component of the exocyst, promotes notch signaling during the asymmetric division of *Drosophila* sensory organ precursors. Dev Cell 9:351–363

Jaleco AC, Neves H, Hooijberg E, Gameiro P, Clode N, Haury M, Henrique D, Parreira L (2001) Differential effects of Notch ligands Delta-1 and Jagged-1 in human lymphoid differentiation. J Exp Med 194:991–1002

Jarriault S, Brou C, Logeat F, Schroeter EH, Kopan R, Israel A (1995) Signalling downstream of activated mammalian Notch. Nature 377:355–358

Jiang R, Lan Y, Chapman HD, Shawber C, Norton CR, Serreze DV, Weinmaster G, Gridley T (1998) Defects in limb, craniofacial, and thymic development in Jagged2 mutant mice. Genes Dev 12:1046–1057

Johnston SH, Rauskolb C, Wilson R, Prabhakaran B, Irvine KD, Vogt TF (1997) A family of mammalian Fringe genes implicated in boundary determination and the Notch pathway. Development 124:2245–2254

Jung J, Mo JS, Kim MY, Ann EJ, Yoon JH, Park HS (2011) Regulation of Notch1 signaling by Delta-like ligand 1 intracellular domain through physical interaction. Mol Cells 32:161–165

Kandachar V, Bai T, Chang HC (2008) The clathrin-binding motif and the J-domain of Drosophila Auxilin are essential for facilitating Notch ligand endocytosis. BMC Dev Biol 8:50

Kao HY, Ordentlich P, Koyano-Nakagawa N, Tang Z, Downes M, Kintner CR, Evans RM, Kadesch T (1998) A histone deacetylase corepressor complex regulates the Notch signal transduction pathway. Genes Dev 12:2269–2277

Klueg KM, Parody TR, Muskavitch MA (1998) Complex proteolytic processing acts on Delta, a transmembrane ligand for Notch, during Drosophila development. Mol Biol Cell 9:1709–1723

Koch U, Lacombe TA, Holland D, Bowman JL, Cohen BL, Egan SE, Guidos CJ (2001) Subversion of the T/B lineage decision in the thymus by lunatic fringe-mediated inhibition of Notch-1. Immunity 15:225–236

Koch U, Fiorini E, Benedito R, Besseyrias V, Schuster-Gossler K, Pierres M, Manley NR, Duarte A, Macdonald HR, Radtke F (2008) Delta-like 4 is the essential, nonredundant ligand for Notch1 during thymic T cell lineage commitment. J Exp Med 205:2515–2523

Komatsu H, Chao MY, Larkins-Ford J, Corkins ME, Somers GA, Tucey T, Dionne HM, White JQ, Wani K, Boxem M, Hart AC (2008) OSM-11 facilitates LIN-12 Notch signaling during *Caenorhabditis elegans* vulval development. PLoS Biol 6:e196

Koo BK, Lim HS, Song R, Yoon MJ, Yoon KJ, Moon JS, Kim YW, Kwon MC, Yoo KW, Kong MP, Lee J, Chitnis AB, Kim CH, Kong YY (2005a) Mind bomb 1 is essential for generating functional Notch ligands to activate Notch. Development 132:3459–3470

Koo BK, Yoon KJ, Yoo KW, Lim HS, Song R, So JH, Kim CH, Kong YY (2005b) Mind bomb-2 is an E3 ligase for Notch ligand. J Biol Chem 280:22335–22342

Koo BK, Yoon MJ, Yoon KJ, Im SK, Kim YY, Kim CH, Suh PG, Jan YN, Kong YY (2007) An obligatory role of mind bomb-1 in notch signaling of mammalian development. PLoS One 2:e1221

Kopan R, Ilagan MX (2009) The canonical Notch signaling pathway: unfolding the activation mechanism. Cell 137:216–233

Kovall RA (2007) Structures of CSL, Notch and Mastermind proteins: piecing together an active transcription complex. Curr Opin Struct Biol 17:117–127

Kuroda K, Han H, Tani S, Tanigaki K, Tun T, Furukawa T, Taniguchi Y, Kurooka H, Hamada Y, Toyokuni S, Honjo T (2003) Regulation of marginal zone B cell development by MINT, a suppressor of Notch/RBP-J signaling pathway. Immunity 18:301–312

Ladi E, Nichols JT, Ge W, Miyamoto A, Yao C, Yang LT, Boulter J, Sun YE, Kintner C, Weinmaster G (2005) The divergent DSL ligand Dll3 does not activate Notch signaling but cell autonomously attenuates signaling induced by other DSL ligands. J Cell Biol 170: 983–992

Lai EC, Deblandre GA, Kintner C, Rubin GM (2001) *Drosophila* neuralized is a ubiquitin ligase that promotes the internalization and degradation of delta. Dev Cell 1:783–794

Lai EC, Roegiers F, Qin X, Jan YN, Rubin GM (2005) The ubiquitin ligase *Drosophila* Mind bomb promotes Notch signaling by regulating the localization and activity of Serrate and Delta. Development 132:2319–2332

Laky K, Fowlkes BJ (2007) Presenilins regulate alphabeta T cell development by modulating TCR signaling. J Exp Med 204:2115–2129

Lamar E, Deblandre G, Wettstein D, Gawantka V, Pollet N, Niehrs C, Kintner C (2001) Nrarp is a novel intracellular component of the Notch signaling pathway. Genes Dev 15:1885–1899

Lardelli M, Lendahl U (1993) Motch A and motch B—two mouse Notch homologues coexpressed in a wide variety of tissues. Exp Cell Res 204:364–372

Lardelli M, Dahlstrand J, Lendahl U (1994) The novel Notch homologue mouse Notch 3 lacks specific epidermal growth factor-repeats and is expressed in proliferating neuroepithelium. Mech Dev 46:123–136

Le Borgne R, Schweisguth F (2003a) Notch signaling: endocytosis makes delta signal better. Curr Biol 13:R273–R275

Le Borgne R, Schweisguth F (2003b) Unequal segregation of Neuralized biases Notch activation during asymmetric cell division. Dev Cell 5:139–148

Lehar SM, Dooley J, Farr AG, Bevan MJ (2005) Notch ligands Delta 1 and Jagged1 transmit distinct signals to T-cell precursors. Blood 105:1440–1447

Lehmann RJF, Dietrich U, Campos-Ortega JA (1983) On the phenotype and development of mutants of early neurogenesis in *Drosophila melanogaster*. Roux's Arch Dev Biol 192:62–74

Li N, Boyd K, Dempsey PJ, Vignali DA (2007) Non-cell autonomous expression of TNF-alpha-converting enzyme ADAM17 is required for normal lymphocyte development. J Immunol 178:4214–4221

Lindsell CE, Shawber CJ, Boulter J, Weinmaster G (1995) Jagged: a mammalian ligand that activates Notch1. Cell 80:909–917

Liu AP, Aguet F, Danuser G, Schmid SL (2010) Local clustering of transferrin receptors promotes clathrin-coated pit initiation. J Cell Biol 191:1381–1393

Maillard I, Weng AP, Carpenter AC, Rodriguez CG, Sai H, Xu L, Allman D, Aster JC, Pear WS (2004) Mastermind critically regulates Notch-mediated lymphoid cell fate decisions. Blood 104:1696–1702

Maillard I, Tu L, Sambandam A, Yashiro-Ohtani Y, Millholland J, Keeshan K, Shestova O, Xu L, Bhandoola A, Pear WS (2006) The requirement for Notch signaling at the beta-selection checkpoint in vivo is absolute and independent of the pre-T cell receptor. J Exp Med 203:2239–2245

Mancini SJ, Mantei N, Dumortier A, Suter U, MacDonald HR, Radtke F (2005) Jagged1-dependent Notch signaling is dispensable for hematopoietic stem cell self-renewal and differentiation. Blood 105:2340–2342

Matsuno K, Ito M, Hori K, Miyashita F, Suzuki S, Kishi N, Artavanis-Tsakonas S, Okano H (2002) Involvement of a proline-rich motif and RING-H2 finger of Deltex in the regulation of Notch signaling. Development 129:1049–1059

McGill MA, McGlade CJ (2003) Mammalian numb proteins promote Notch1 receptor ubiquitination and degradation of the Notch1 intracellular domain. J Biol Chem 278:23196–23203

McMahon HT, Gallop JL (2005) Membrane curvature and mechanisms of dynamic cell membrane remodelling. Nature 438:590–596

Michie AM, Zuniga-Pflucker JC (2002) Regulation of thymocyte differentiation: pre-TCR signals and beta-selection. Semin Immunol 14:311–323

Mishra-Gorur K, Rand MD, Perez-Villamil B, Artavanis-Tsakonas S (2002) Down-regulation of Delta by proteolytic processing. J Cell Biol 159:313–324

Mohtashami M, Shah DK, Nakase H, Kianizad K, Petrie HT, Zuniga-Pflucker JC (2010) Direct comparison of Dll1- and Dll4-mediated Notch activation levels shows differential lymphomyeloid lineage commitment outcomes. J Immunol 185:867–876

Morgan TH (1917) The theory of the gene. Yale University Press, New Haven

Nichols JT, Miyamoto A, Olsen SL, D'Souza B, Yao C, Weinmaster G (2007) DSL ligand endocytosis physically dissociates Notch1 heterodimers before activating proteolysis can occur. J Cell Biol 176:445–458

Oberg C, Li J, Pauley A, Wolf E, Gurney M, Lendahl U (2001) The Notch intracellular domain is ubiquitinated and negatively regulated by the mammalian Sel-10 homolog. J Biol Chem 276:35847–35853

Okajima T, Irvine KD (2002) Regulation of notch signaling by o-linked fucose. Cell 111:893–904

Oswald F, Winkler M, Cao Y, Astrahantseff K, Bourteele S, Knochel W, Borggrefe T (2005) RBP-Jkappa/SHARP recruits CtIP/CtBP corepressors to silence Notch target genes. Mol Cell Biol 25:10379–10390

Overstreet E, Fitch E, Fischer JA (2004) Fat facets and Liquid facets promote Delta endocytosis and Delta signaling in the signaling cells. Development 131:5355–5366

Panin VM, Papayannopoulos V, Wilson R, Irvine KD (1997) Fringe modulates Notch-ligand interactions. Nature 387:908–912

Parks AL, Klueg KM, Stout JR, Muskavitch MA (2000) Ligand endocytosis drives receptor dissociation and activation in the Notch pathway. Development 127:1373–1385

Parks AL, Stout JR, Shepard SB, Klueg KM, Dos Santos AA, Parody TR, Vaskova M, Muskavitch MA (2006) Structure-function analysis of delta trafficking, receptor binding and signaling in *Drosophila*. Genetics 174:1947–1961

Pavlopoulos E, Pitsouli C, Klueg KM, Muskavitch MA, Moschonas NK, Delidakis C (2001) neuralized Encodes a peripheral membrane protein involved in delta signaling and endocytosis. Dev Cell 1:807–816

Petrie HT, Zuniga-Pflucker JC (2007) Zoned out: functional mapping of stromal signaling microenvironments in the thymus. Annu Rev Immunol 25:649–679

Pintar A, De Biasio A, Popovic M, Ivanova N, Pongor S (2007) The intracellular region of Notch ligands: does the tail make the difference? Biol Direct 2:19

Pitsouli C, Delidakis C (2005) The interplay between DSL proteins and ubiquitin ligases in Notch signaling. Development 132:4041–4050

Pui JC, Allman D, Xu L, DeRocco S, Karnell FG, Bakkour S, Lee JY, Kadesch T, Hardy RR, Aster JC, Pear WS (1999) Notch1 expression in early lymphopoiesis influences B versus T lineage determination. Immunity 11:299–308

Radtke F, Wilson A, Stark G, Bauer M, van Meerwijk J, MacDonald HR, Aguet M (1999) Deficient T cell fate specification in mice with an induced inactivation of Notch1. Immunity 10:547–558

Rebay I, Fleming RJ, Fehon RG, Cherbas L, Cherbas P, Artavanis-Tsakonas S (1991) Specific EGF repeats of Notch mediate interactions with Delta and Serrate: implications for Notch as a multifunctional receptor. Cell 67:687–699

Reizis B, Leder P (2002) Direct induction of T lymphocyte-specific gene expression by the mammalian Notch signaling pathway. Genes Dev 16:295–300

Rhyu MS, Jan LY, Jan YN (1994) Asymmetric distribution of numb protein during division of the sensory organ precursor cell confers distinct fates to daughter cells. Cell 76:477–491

Robey E, Chang D, Itano A, Cado D, Alexander H, Lans D, Weinmaster G, Salmon P (1996) An activated form of Notch influences the choice between CD4 and CD8 T cell lineages. Cell 87:483–492

Ruan Y, Tecott L, Jiang MM, Jan LY, Jan YN (2001) Ethanol hypersensitivity and olfactory discrimination defect in mice lacking a homolog of *Drosophila* neuralized. Proc Natl Acad Sci U S A 98:9907–9912

Sasamura T, Sasaki N, Miyashita F, Nakao S, Ishikawa HO, Ito M, Kitagawa M, Harigaya K, Spana E, Bilder D, Perrimon N, Matsuno K (2003) neurotic, a novel maternal neurogenic gene, encodes an O-fucosyltransferase that is essential for Notch-Delta interactions. Development 130:4785–4795

Schmitt TM, Zuniga-Pflucker JC (2002) Induction of T cell development from hematopoietic progenitor cells by delta-like-1 in vitro. Immunity 17:749–756

Schmitt TM, Ciofani M, Petrie HT, Zuniga-Pflucker JC (2004) Maintenance of T cell specification and differentiation requires recurrent notch receptor-ligand interactions. J Exp Med 200:469–479

Seugnet L, Simpson P, Haenlin M (1997) Requirement for dynamin during Notch signaling in *Drosophila* neurogenesis. Dev Biol 192:585–598

Shawber C, Boulter J, Lindsell CE, Weinmaster G (1996) Jagged2: a serrate-like gene expressed during rat embryogenesis. Dev Biol 180:370–376

Shi S, Stanley P (2003) Protein O-fucosyltransferase 1 is an essential component of Notch signaling pathways. Proc Natl Acad Sci U S A 100:5234–5239

Shi J, Fallahi M, Luo JL, Petrie HT (2011) Nonoverlapping functions for Notch1 and Notch3 during murine steady-state thymic lymphopoiesis. Blood 118:2511–2519

Shimizu K, Chiba S, Kumano K, Hosoya N, Takahashi T, Kanda Y, Hamada Y, Yazaki Y, Hirai H (1999) Mouse jagged1 physically interacts with notch2 and other notch receptors. Assessment by quantitative methods. J Biol Chem 274:32961–32969

Shutter JR, Scully S, Fan W, Richards WG, Kitajewski J, Deblandre GA, Kintner CR, Stark KL (2000) Dll4, a novel Notch ligand expressed in arterial endothelium. Genes Dev 14:1313–1318

Skwarek LC, Garroni MK, Commisso C, Boulianne GL (2007) Neuralized contains a phosphoinositide-binding motif required downstream of ubiquitination for delta endocytosis and notch signaling. Dev Cell 13:783–795

Song R, Koo BK, Yoon KJ, Yoon MJ, Yoo KW, Kim HT, Oh HJ, Kim YY, Han JK, Kim CH, Kong YY (2006) Neuralized-2 regulates a Notch ligand in cooperation with Mind bomb-1. J Biol Chem 281:36391–36400

Song R, Kim YW, Koo BK, Jeong HW, Yoon MJ, Yoon KJ, Jun DJ, Im SK, Shin J, Kong MP, Kim KT, Yoon K, Kong YY (2008) Mind bomb 1 in the lymphopoietic niches is essential for T and marginal zone B cell development. J Exp Med 205:2525–2536

Spana EP, Doe CQ (1996) Numb antagonizes Notch signaling to specify sibling neuron cell fates. Neuron 17:21–26

Stanley P, Okajima T (2010) Roles of glycosylation in Notch signaling. Curr Top Dev Biol 92:131–164

Staub O, Rotin D (2006) Role of ubiquitylation in cellular membrane transport. Physiol Rev 86:669–707

Su DM, Navarre S, Oh WJ, Condie BG, Manley NR (2003) A domain of Foxn1 required for crosstalk-dependent thymic epithelial cell differentiation. Nat Immunol 4:1128–1135

Suliman S, Tan J, Xu K, Kousis PC, Kowalski PE, Chang G, Egan SE, Guidos C (2011) Notch3 is dispensable for thymocyte beta-selection and Notch1-induced T cell leukemogenesis. PLoS ONE 6:e24937

Takeuchi T, Adachi Y, Ohtsuki Y (2005) Skeletrophin, a novel RING molecule controlled by the chromatin remodeling complex, is downregulated in malignant melanoma. DNA Cell Biol 24:339–344

Tamura K, Taniguchi Y, Minoguchi S, Sakai T, Tun T, Furukawa T, Honjo T (1995) Physical interaction between a novel domain of the receptor Notch and the transcription factor RBP-J kappa/Su(H). Curr Biol 5:1416–1423

Tanigaki K, Tsuji M, Yamamoto N, Han H, Tsukada J, Inoue H, Kubo M, Honjo T (2004) Regulation of alphabeta/gammadelta T cell lineage commitment and peripheral T cell responses by Notch/RBP-J signaling. Immunity 20:611–622

Thompson PK, Zuniga-Pflucker JC (2011) On becoming a T cell, a convergence of factors kick it up a Notch along the way. Semin Immunol 23:350–359

Tian X, Hansen D, Schedl T, Skeath JB (2004) Epsin potentiates Notch pathway activity in *Drosophila* and *C. elegans*. Development 131:5807–5815
Tian L, Wu X, Chi C, Han M, Xu T, Zhuang Y (2008) ADAM10 is essential for proteolytic activation of Notch during thymocyte development. Int Immunol 20:1181–1187
Tien AC, Rajan A, Bellen HJ (2009) A Notch updated. J Cell Biol 184:621–629
Uemura T, Shepherd S, Ackerman L, Jan LY, Jan YN (1989) numb, a gene required in determination of cell fate during sensory organ formation in *Drosophila* embryos. Cell 58:349–360
Uyttendaele H, Marazzi G, Wu G, Yan Q, Sassoon D, Kitajewski J (1996) Notch4/int-3, a mammary proto-oncogene, is an endothelial cell-specific mammalian Notch gene. Development 122:2251–2259
Van de Walle I, De Smet G, Gartner M, De Smedt M, Waegemans E, Vandekerckhove B, Leclercq G, Plum J, Aster JC, Bernstein ID, Guidos CJ, Kyewski B, Taghon T (2011) Jagged2 acts as a Delta-like Notch ligand during early hematopoietic cell fate decisions. Blood 117:4449–4459
Varas A, Hager-Theodorides AL, Sacedon R, Vicente A, Zapata AG, Crompton T (2003) The role of morphogens in T-cell development. Trends Immunol 24:197–206
Varnum-Finney B, Wu L, Yu M, Brashem-Stein C, Staats S, Flowers D, Griffin JD, Bernstein ID (2000) Immobilization of Notch ligand, Delta-1, is required for induction of notch signaling. J Cell Sci 113(Pt 23):4313–4318
Vitt UA, Hsu SY, Hsueh AJ (2001) Evolution and classification of cystine knot-containing hormones and related extracellular signaling molecules. Mol Endocrinol 15:681–694
Vollrath B, Pudney J, Asa S, Leder P, Fitzgerald K (2001) Isolation of a murine homologue of the *Drosophila* neuralized gene, a gene required for axonemal integrity in spermatozoa and terminal maturation of the mammary gland. Mol Cell Biol 21:7481–7494
Wang W, Struhl G (2004) *Drosophila* Epsin mediates a select endocytic pathway that DSL ligands must enter to activate Notch. Development 131:5367–5380
Washburn T, Schweighoffer E, Gridley T, Chang D, Fowlkes BJ, Cado D, Robey E (1997) Notch activity influences the alphabeta versus gammadelta T cell lineage decision. Cell 88:833–843
Weinmaster G, Roberts VJ, Lemke G (1992) Notch2: a second mammalian Notch gene. Development 116:931–941
Wilson A, Ardiet DL, Saner C, Vilain N, Beermann F, Aguet M, Macdonald HR, Zilian O (2007) Normal hemopoiesis and lymphopoiesis in the combined absence of numb and numblike. J Immunol 178:6746–6751
Windler SL, Bilder D (2010) Endocytic internalization routes required for delta/notch signaling. Curr Biol 20:538–543
Wolfer A, Bakker T, Wilson A, Nicolas M, Ioannidis V, Littman DR, Lee PP, Wilson CB, Held W, MacDonald HR, Radtke F (2001) Inactivation of Notch 1 in immature thymocytes does not perturb CD4 or CD8T cell development. Nat Immunol 2:235–241
Wolfer A, Wilson A, Nemir M, MacDonald HR, Radtke F (2002) Inactivation of Notch1 impairs VDJbeta rearrangement and allows pre-TCR-independent survival of early alpha beta Lineage Thymocytes. Immunity 16:869–879
Wong GW, Zuniga-Pflucker JC (2010) gammadelta and alphabeta T cell lineage choice: resolution by a stronger sense of being. Semin Immunol 22:228–236
Wu L, Aster JC, Blacklow SC, Lake R, Artavanis-Tsakonas S, Griffin JD (2000) MAML1, a human homologue of *Drosophila* mastermind, is a transcriptional co-activator for NOTCH receptors. Nat Genet 26:484–489
Wu L, Sun T, Kobayashi K, Gao P, Griffin JD (2002) Identification of a family of mastermind-like transcriptional coactivators for mammalian notch receptors. Mol Cell Biol 22:7688–7700
Xiao S, Su DM, Manley NR (2008) T cell development from kit-negative progenitors in the Foxn1Delta/Delta mutant thymus. J Immunol 180:914–921
Yamamoto S, Charng WL, Bellen HJ (2010) Endocytosis and intracellular trafficking of Notch and its ligands. Curr Top Dev Biol 92:165–200
Yeh E, Dermer M, Commisso C, Zhou L, McGlade CJ, Boulianne GL (2001) Neuralized functions as an E3 ubiquitin ligase during *Drosophila* development. Curr Biol 11:1675–1679

Yuan JS, Tan JB, Visan I, Matei IR, Urbanellis P, Xu K, Danska JS, Egan SE, Guidos CJ (2011) Lunatic Fringe prolongs Delta/Notch-induced self-renewal of committed alphabeta T-cell progenitors. Blood 117:1184–1195

Zhang C, Li Q, Lim CH, Qiu X, Jiang YJ (2007) The characterization of zebrafish antimorphic mib alleles reveals that Mib and Mind bomb-2 (Mib2) function redundantly. Dev Biol 305:14–27

Zhang L, Widau RC, Herring BP, Gallagher PJ (2011) Delta-like 1-Lysine613 regulates notch signaling. Biochim Biophys Acta 1813:2036–2043

Zhong W, Jiang MM, Weinmaster G, Jan LY, Jan YN (1997) Differential expression of mammalian Numb, Numblike and Notch1 suggests distinct roles during mouse cortical neurogenesis. Development 124:1887–1897

Zuniga-Pflucker JC (2004) T-cell development made simple. Nat Rev Immunol 4:67–72

Notch1 and IL-7 Receptor Signalling in Early T-cell Development and Leukaemia

Sara González-García, Marina García-Peydró, Juan Alcain and María L. Toribio

Abstract Notch receptors are master regulators of many aspects of development and tissue renewal in metazoans. Notch1 activation is essential for T-cell specification of bone marrow-derived multipotent progenitors that seed the thymus, and for proliferation and further progression of early thymocytes along the T-cell lineage. Deregulated activation of Notch1 significantly contributes to the generation of T-cell acute lymphoblastic leukaemia (T-ALL). In addition to Notch1 signals, survival and proliferation signals provided by the IL-7 receptor (IL-7R) are also required during thymopoiesis. Our understanding of the molecular mechanisms controlling stage-specific survival and proliferation signals provided by Notch1 and IL-7R has recently been improved by the discovery that the IL-7R is a transcriptional target of Notch1. Thus, Notch1 controls T-cell development, in part by regulating the stage- and lineage-specific expression of IL-7R. The finding that induction of IL-7R expression downstream of Notch1 also occurs in T-ALL highlights the important contribution that deregulated IL-7R expression and function may have in this pathology. Confirming this notion, oncogenic *IL7R* gain-of-function mutations have recently been identified in childhood T-ALL. Here we discuss the fundamental role of Notch1 and IL-7R signalling pathways in physiological and pathological T-cell development in mice and men, highlighting their close molecular underpinnings.

Keywords Thymus · T-cell lineage · Notch · IL-7R · lympho-myeloid progenitor · leukaemia

S. González-García · M. García-Peydró · J. Alcain
Centro de Biología Molecular "Severo Ochoa" Consejo Superior de Investigaciones Científicas, Universidad Autónoma de Madrid, 28049, Madrid, Spain

M. L. Toribio (✉)
Centro de Biología Molecular "Severo Ochoa", Universidad Autónoma de Madrid, C/Nicolás Cabrera1, 28049, Madrid, Spain
e-mail: mtoribio@cbm.uam.es

Current Topics in Microbiology and Immunology (2012) 360: 47–73
DOI: 10.1007/82_2012_231

Published Online: 14 June 2012

Contents

1 From Early Thymic Progenitors to Mature T Cells

All blood lineages but T cells derive in situ within the bone marrow (BM) from resident haematopoietic stem cells (HSCs) that undergo a maturation process characterised by the progressive loss of developmental potentials and the activation of lineage-specific transcriptional programs. T lymphocyte development is an exception, as it occurs within a dedicated lymphoid organ, the thymus. T cells develop from BM-derived progenitors that have lost self-renewing capacity and are thus distinct from HSCs, but still display multi-lineage potential (Fowlkes 1985; Shortman and Wu, 1996; Weerkamp et al. 2006; Blom and Spits 2006). Within the thymus, these haematopoietic progenitors undergo tightly regulated proliferation, maturation and selection processes that require their dynamic relocation within specific niches of the thymic microenvironment (Takahama 2006; Ciofani and Zúñiga-Pflücker 2007; Bhandoola et al. 2007; Rothenberg 2007; Anderson et al. 2007; Petrie and Zúñiga-Pflücker 2007). Both surface molecules and soluble factors from thymic stromal cells, mainly thymic epithelial cells (TECs), provide inductive signals that are essential for directing thymic immigrants along the T-cell linage. Specifically, interactions of Notch receptors on thymic precursors with their ligands expressed on TECs, and signalling mediated by the interleukin-7 receptor (IL-7R) in response to TEC-derived IL-7 (Fig. 1), are crucial events that regulate thymopoiesis in both mouse and man (Shortman et al. 1990; Blom and Spits 2006; Maillard et al. 2005). Deregulation of any of these pathways is, thus, linked to the emergence of T-cell pathologies, largely T-cell immunodeficiencies and leukaemias. The details of how Notch and IL-7R signalling control physiological and leukaemic T-cell development in humans represent the scope of this review.

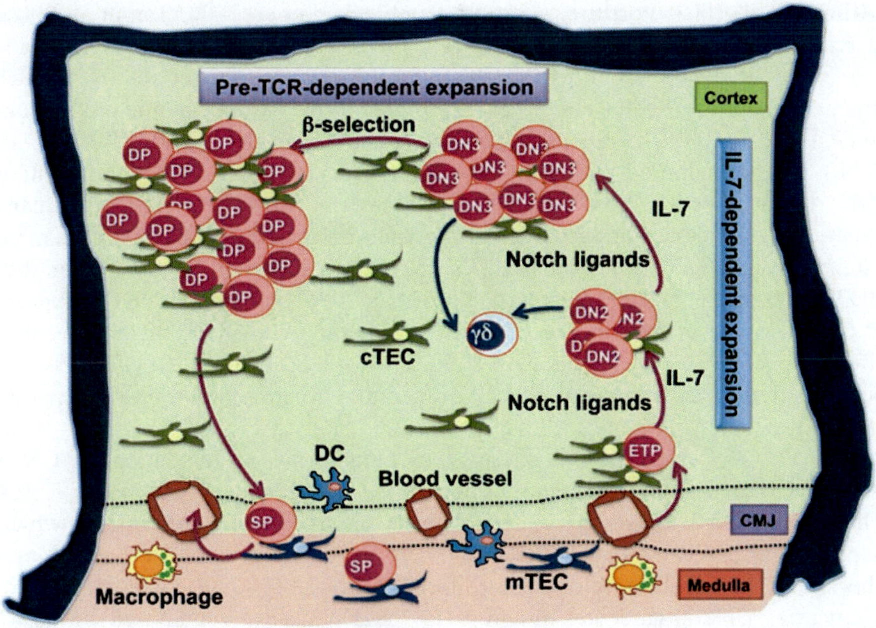

Fig. 1 *Overview of thymic T-cell development.* Bonem arrow-derived lympho-myeloid progenitors enter the thymus through blood vessels located at the cortico–medullary junction (CMJ) and give rise to the pool of (ETPs). Subsequently, ETPs initiate migration towards the external thymic cortex and interact with Notch ligands expressed on cTECs by means of Notch1 expression. Notch1 signalling induces T-cell commitment, IL-7R expression and progression through the $CD4^{-}$ $CD8^{-}$ DN2 and DN3 stages. Interaction of IL-7R on developing thymocytes with locally produced levels of IL-7 triggers a first wave of expansion of the T-cell progenitor pool. DN2 or DN3 thymocytes that succeed in TCRγ and TCRδ rearrangements give rise to TCR$\gamma\delta$ T cells, while successful rearrangement at the TCRβ locus enables the expression of a TCRβ-pTα pre-TCR complex at the DN3 stage. PreTCR signalling promotes cell survival and extensive proliferation as well as progression to the $CD4^{+}$ $CD8^{+}$ DP stage. This process, known as β-selection, represents the second checkpoint of intrathymic expansion in T-cell development. At this stage, proliferation stops and DP thymocytes undergo rearrangement at the TCRα locus and express the TCR$\alpha\beta$. TCR$\alpha\beta$-expressing DP thymocytes then undergo positive and negative selection processes by means of interactions with cTECs or medullary epithelial cells (mTECs) and DCs, respectively. Selected cells exit the thymus through blood vessels at the CMJ as $CD4^{+}$ or $CD8^{+}$ SP thymocytes to establish the peripheral T-cell pool

1.1 Two Intrathymic Checkpoints for Cellular Expansion of Developing T Cells

The particular identity of the thymic immigrants has been elusive owing to the rarity of these cells. Extensive studies lend support to the current notion that in both mouse and man the thymus is seeded by haematopoietic progenitors with a distinctive lineage marker-negative (Lin^{-}) $ckit^{+}$ $Flt3^{+}$ IL-7$R^{-/lo}$ phenotype

(Allman et al. 2003; Porritt et al. 2004; Sambandam et al. 2005; Tan et al. 2005; García-Peydró et al. 2006; Blom and Spits 2006), equivalent to that of lymphoid-primed multipotent progenitors (LMPPs) in the BM (Adolfsson et al. 2005). After thymus arrival, these cells, commonly referred to as the earliest thymic progenitors (ETPs) (0.01–0.03 % of neonatal thymocytes) (Allman et al. 2003; Benz et al. 2008), can be identified within the most immature $CD4^{-}$ $CD8^{-}$ double negative (DN) thymocytes (termed DN1), characterised as $CD44^{+}$ $CD34^{+}$ in humans (Márquez et al. 1995; Carrasco et al. 2002; Spits 2002) and $CD44^{+}$ $CD25^{-}$ in mice (Shortman and Wu 1996). Human ETPs express low levels of the myeloid marker CD33 (Márquez et al. 1998; García-Peydró et al. 2006), while mouse ETPs express PSGL1 and the chemokine receptor CCR9, both involved in thymus seeding (Benz and Bleul 2005; Gossens et al. 2009; Love and Bhandoola 2011). ETPs have been referred to as "canonical" T-cell progenitors because they robustly generate DN2 downstream progeny (Allman et al. 2003; Porritt et al. 2004; Benz et al. 2008) (Fig. 2), which is characterised in humans by up-regulation of CD1a ($CD44^{+}$ $CD25^{+}$ in mice) and commitment to the T-cell lineage (Galy et al. 1993). The ETP to DN2 transition is marked by a massive cellular expansion, which depends on IL-7R-mediated signalling (Fig. 1) and represents a fundamental checkpoint in T-cell development (Shortman et al. 1990; Plum et al. 1996). Thereafter, DN2 thymocytes progress to the DN3 stage, composed of $CD4^{+}$ immature single positive (ISP) thymocytes in humans (Kraft et al. 1993) ($CD44^{-}$ $CD25^{+}$ in mice). At this stage, proliferation stops and gene rearrangement at the TCRδ, γ and β loci is completed (Ramiro et al. 1996; Blom et al. 1999). DN thymocytes that succeed in TCRγ and TCRδ rearrangements will give rise to TCR$\gamma\delta$ T cells, while those expressing a functional TCRβ chain will accomplish progression beyond the DN3 stage by signalling through a highly conserved TCRβ–pTα (pre-TCR) complex (von Boehmer and Fehling 1997; Ramiro et al. 1996; Carrasco et al. 1999, 2001). Cell surface pre-TCR expression promotes cell survival and proliferation and induces progression to the $CD4^{+}$ $CD8^{+}$ double positive (DP) stage by a process known as β-selection (von Boehmer et al. 1998), which represents the second checkpoint of intrathymic expansion in T-cell development (Fig. 1). Thereafter, the pre-TCR is down-regulated and DP thymocytes stop cell division, undergo rearrangements at the TCRα locus (von Boehmer and Fehling 1997; Trigueros et al. 1998; Carrasco et al. 2001) and finally express a mature TCR$\alpha\beta$ that will allow their positive selection and differentiation into mature $CD4^{+}$ or $CD8^{+}$ single positive (SP) thymocytes. SP thymocytes that survive negative selection migrate to the periphery as MHC-restricted self-tolerant T-cells (Takahama 2006; Anderson et al. 2007). Therefore, thymopoiesis encompasses two checkpoints of cellular expansion sequentially controlled by the IL-7R and the pre-TCR, whose expression has to be tightly regulated to ensure physiological proliferation of developing thymocytes.

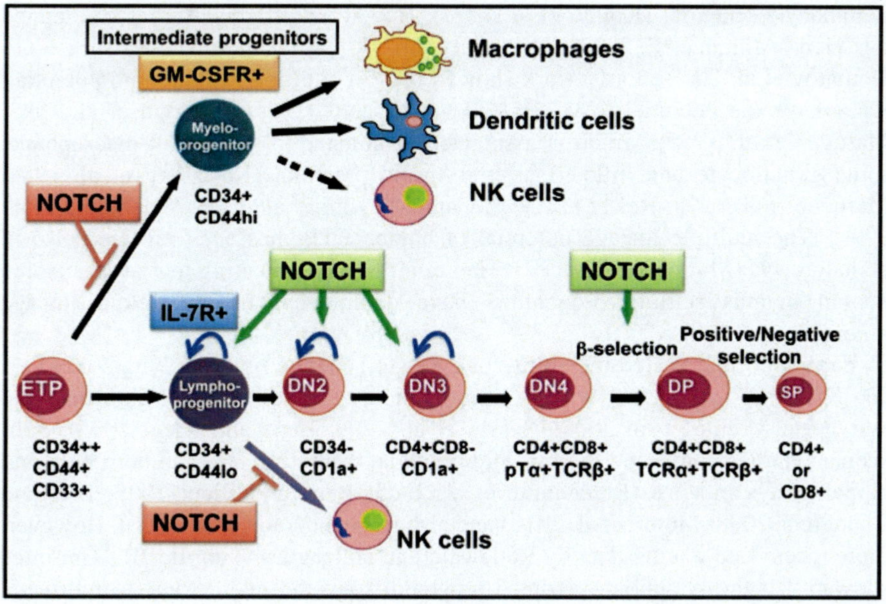

Fig. 2 *Proposed model of lymphoid and myeloid developmet in human thymus.* Human ETPs are lympho-myeloid precursors that can generate separate lymphoid- or myeloid-primed intermediate progenitors, characterised by the selective expression of either the IL-7R or the GM–CSFR, respectively. Lymphoid progenitors display T- and NK-cell potentials, while myeloid progenitors have lost T-cell potential, retain some NK-cell potential, and are enriched in myeloid (macrophage and DCs) potentials. Notch1 signalling diverts ETPs away from the myeloid pathway by blocking the generation of myeloid-primed progenitors, thus enforcing the lymphoid cell fate. Simultaneously, Notch1 induces the expression of IL-7R, which marks the lymphoid–myeloid branching point and triggers proliferation of lymphoid-primed progenitors. Thereafter, sustained Notch1 signalling blocks NK development from bipotent T/NK lymphoid progenitors and favours development along the T-cell lineage. This process is characterised by the progressive loss of CD34, CD44 and CD33 and the sequential acquisition of CD1a by $CD4^-$ $CD8^-$ DN thymocytes (DN2 stage) and CD4 molecules (DN3 stage). Successful TCRβ gene rearrangement at the DN3 stage results in TCRβ association with the invariant pTα chain and formation of a pre-TCR complex at the DN4 stage. The pre-TCR cooperates with Notch1 signals at the β-selection checkpoint to induce a second wave of intrathymic proliferation and progression to the DP stage. Further rearrangement at the TCRα locus in DP thymocytes enables the expression of a mature TCR$\alpha\beta$. DP cells that survive TCR$\alpha\beta$-mediated positive and negative selection give rise to $CD4^+$ or $CD8^+$ SP thymocytes that migrate to the periphery

1.2 IL-7R Expression Marks a Lymphoid/Myeloid Developmental Split in Human Thymopoiesis

Understanding how ETPs differentiate along the T-cell lineage has been challenging owing to discrepancies about their level of commitment and lineage potential. The so-called classical model proposed that the thymus is colonised exclusively by common lymphoid progenitors (CLP) harbouring T, NK and B lymphocyte potential (Kondo et al. 1997). However, different studies challenged this view (Allman et al. 2003; Adolfsson et al. 2005; Kawamoto and Katsura 2009; Doulatov et al. 2010). Early work showed that, in addition to lymphoid potential, ETPs have the potential to generate dendritic cells (DCs) (Ardavin et al. 1993; Márquez et al. 1995), which is associated in humans to significant macrophage, some granulocyte and little, if any, erythroid potential (Kurtzberg et al. 1989; Márquez et al. 1995; Res et al. 1996; García-Peydró et al. 2006; Weerkamp et al. 2006). The multiple lineage potential of human ETPs has been revealed also in clonal assays (Márquez et al. 1998; Hao et al. 2008), indicating that at least some human thymus settling progenitors have combined lympho-myeloid lineage potential.

Early reports revealed the ability of ETPs to generate multiple lymphoid (T, B and NK) lineages together with DCs in the mouse (Wu et al. 1991; Shortman and Wu 1996; Michie et al. 2000; Akashi et al. 2000; Benz and Bleul 2005). Subsequent studies confirmed the myeloid potential of murine ETPs in both bulk and clonal assays in vitro (Balciunaite et al. 2005; Benz and Bleul 2005; Bell and Bhandoola 2008; Moore et al. 2012) and also in vivo (Wada et al. 2008). However, more recent fate tracing data by Rodewald and colleagues using IL-7Rα-Cre mice showed that the lymphoid-restricted progenitors are the major route to murine T cells in vivo (Schlenner et al. 2010), suggesting a separate origin of T cells and myeloid cells in the murine thymus. Although the field remains controversial, functional and molecular evidence has lately been provided at the single cell level that the earliest progenitors in the neonatal murine thymus, as well as LMPPs in the BM, have combined granulocyte–macrophage, T lymphocyte and B-cell lymphopoietic potentials, but not megakaryocyte–erythroid lineage potential (Luc et al. 2012; Ceredig 2012), as reported for human ETPs (Weerkamp et al. 2006; García-Peydró et al. 2006).

Realization of the lympho-myeloid potential of ETPs is of crucial relevance for understanding how either T-cell or alternative cell fates are imposed within the thymus. Several studies support that T-cell commitment in mice is a sequential process, involving progressively increasing limitations on the non-T cell potential of ETPs. Loss of B-cell potential may happen immediately after thymus entry or even prior to thymus seeding (Benz and Bleul 2005; Harman et al. 2005; Heinzel et al. 2007), and is followed by loss of myeloid, DC and NK cell potentials (Ciofani and Zúñiga-Pflücker 2007; Bhandoola et al. 2007; Rothenberg et al. 2008; Ceredig 2012). In humans, however, our studies suggest that after loss of B-cell potential progression towards the T-lineage fate involves an early, split of ETPs into two, alternative developmental pathways that proceed through independent myeloid- or lymphoid-primed intermediate progenitors (Fig. 2). This branching point is marked both in vitro and in vivo by down-regulation of CD34 and either the up- or down-regulation of CD44 (Márquez et al. 1995). CD44 up-regulation results, in turn, in up-regulated expression of the myeloid marker CD33 and loss of T-cell potential. These non-T cell progenitors are, however, enriched in myeloid-DC potential and retain some NK cell potential. Conversely, CD44 down-regulation parallels an increased expression of the T-cell marker CD5, loss of myeloid-

DC capability and progression towards the T/NK bi-potential lymphoid stage (Márquez et al. 1998; de Yébenes et al. 2002). Expression of CD1a in lymphoid-restricted thymocytes finally marks loss of NK lymphopoietic ability and T-cell commitment (Galy et al. 1993; Sánchez et al. 1994; de Yébenes et al. 2002) (Fig. 2). According to their restricted lineage potentials, intermediate thymic progenitors display exclusive myeloid or lymphoid-associated gene transcription patterns (Martín-Gayo et al. unpublished results). Likewise, myeloid-specific transcriptional priming has recently been reported for myeloid precursors of DCs recently identified in the mouse thymus (Moore et al. 2012).

The myeloid versus lymphoid cell fate decision in human thymus robustly associates with a selective expression of receptors for either granulocyte–macrophage colony-stimulating factor (GM-CSF) or IL-7, respectively, reflecting the cytokine-specific dependence of each developmental pathway (Márquez et al. 1998; de Yébenes et al. 2002, García-Peydró et al. 2006). Therefore, up-regulation of IL-7R marks a key lymphoid-myeloid branching point in human thymopoiesis, which associates with loss of myeloid (macrophage and DC) potentials and with T-lineage specification (Fig. 2).

2 IL-7R Signalling in Early T-cell Development

Different studies have highlighted the conserved non-redundant role played by IL-7 in T-cell development. IL-7 is a cytokine produced by thymic epithelium and BM stromal cells. Binding of IL-7 to the IL-7R, which is composed of the specific α-chain (IL-7Rα) and the common cytokine receptor γ (γc) chain (Fig. 3), results in α and γ chain dimerisation. The, thereby, activated IL-7Rα-γc heterodimer transmits signals for survival and proliferation to T- and also B-lineage cells (Shortman et al. 1990; Akashi et al. 1998; Leonard 2001). Pathways activated downstream of IL-7R include JAK/STAT, PI3K, MAPK and Src kinases, ultimately leading to the expression of target genes such Bcl-2 family members, cyclin D1, SOCS-1 and c-myc (Leonard 2001). Mice deficient in IL-7 or IL-7R have an early block in T-cell development associated with reduced numbers of non-functional T cells and impaired B lymphopoiesis (Sudo et al. 1993; Peschon et al. 1994; von Freeden-Jeffry et al. 1995). Bcl-2 is sufficient to rescue T-cell but not B-cell development in those mice (Maraskovsky et al. 1997, 1998; Akashi et al. 1997), indicating that IL-7 has a pro-survival effect in T-lineage cells. In humans, IL-7R paucity does not affect B lymphopoiesis, but results in severe combined immunodeficiency (SCID) characterised by a complete lack of T cells (Puel et al. 1998; Leonard 2001). Therefore, species-specific lineage-associated IL-7/IL-7R requirements and/or signalling pathways downstream of IL-7R are likely to be responsible for this difference (Pallard et al. 1999).

These results point toward an essential function of IL-7/IL-7R during early T-cell development that supports the survival and the enormous expansion that experiences the intrathymic pool of T-cell-specified DN2 progenitors (Shortman

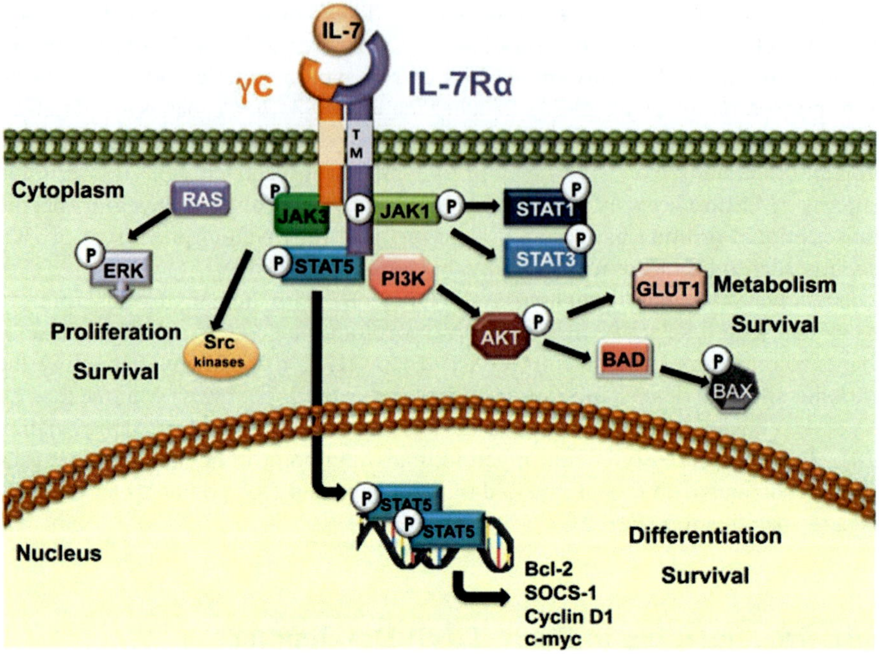

Fig. 3 *IL-7 receptor signalling pathway*. IL-7 receptor α chain (IL-7Rα) and γ-common (γc) chains components of the IL-7R undergo dimerisation upon IL-7 binding. Thereby, the IL-7R heterodimer triggers activation of JAK3 and JAK1 kinases bound to the intracellular domain of γc or IL-7Rα, respectively. JAK1 induces phosphorylation of the IL-7Rα intracellular domain and promotes recruitment of PI3K and STAT proteins. Phosphorylation of STAT proteins by JAK results in their dimerisation and translocation to the nucleus and finally in the transcription of target genes including Bcl-2, SOCS-1, CyclinD1 and c-myc. In addition, PI3K recruited to the intracellular domain of IL-7Rα becomes activated and phosphorylates Akt, which thereafter promotes cell survival through degradation of pro-apoptotic proteins such as Bad and Bax and glucose uptake through the expression of GLUT1. IL-7R signalling also involves activation of the ERK pathway

et al. 1990; Peschon et al. 1994; Plum et al. 1996; Akashi et al. 1998). In addition, signalling through IL-7R at the DN2 stage controls TCRγ locus accessibility and gene rearrangement (Durum et al. 1998). Accordingly, DN2 cells with higher IL-7R expression levels are biased to the γδ T-cell lineage, at least in mice (Kang et al. 2001). IL-7 is, however, dispensable at or beyond the DN3 stage, although it may be required later on during positive selection of $CD8^+$ thymocytes (Munitic et al. 2004; Yu et al. 2004; Van De Wiele et al. 2004). The stage-specific effect of IL-7 is accomplished by a dynamic regulation of IL-7R expression and function (Munitic et al. 2004). Of the two IL-7R components, the γc is constitutively expressed in thymus-seeding precursors, which have negative or low levels of IL-7Rα expression, indicating that IL-7Rα is up-regulated at the ETP stage (García-Peydró et al. 2006; Porritt et al. 2004; Allman et al. 2003). Subsequently, IL-7Rα expression increases progressively until the DN2 stage, declines steadily thereafter, and must be terminated between the β-selection and positive selection checkpoints (Munitic et al. 2004; Yu et al. 2004, 2006; Van De Wiele et al. 2004). Besides transcriptional regulation, active suppression of IL-7R signalling induced by suppressor of cytokine signalling (SOCS)-1 guarantees IL-7R-unresponsiveness in pre-selected DP thymocytes (Yu et al. 2006), which is obligatory to allow negative selection and death by neglect to occur at the DP stage. In fact, sustained IL-7R expression in *il7ra* transgenic mice results in a direct competition between DN progenitors and DP thymocytes for the limiting amounts of endogenous IL-7 within the thymus (Munitic et al. 2004). Finally, restoration of IL-7R surface levels by positive selection will ensure homeostatic proliferation of peripheral mature T cells derived from CD4 and CD8 SP thymocytes (Schluns et al. 2000). Tight regulation of IL-7Rα expression during thymopoiesis, thus, guarantees specific survival and expansion of those intrathymic progenitors that have undergone T-cell specification.

3 Notch Regulation of Early Thymocyte Development

T-cell specification signals uniquely provided by the thymus microenvironment largely rely on the Notch signalling pathway (Takahama 2006; Petrie and Zúñiga-Pflücker 2007; Koch and Radtke 2011). Notch activation is required at sequential intrathymic stages to provide survival, proliferation or metabolic cues as well as differentiation signals, all necessary for efficient T-cell development (Schmitt et al. 2004; Ciofani and Zúñiga-Pflücker 2007). The Notch pathway is an evolutionary conserved cell signalling system of transmembrane receptors that includes four members in mammals (Notch1–4). Notch1 is the receptor preferentially expressed in developing thymocytes, although all members are expressed in the thymus (Fiorini et al. 2009; Van de Walle et al. 2011; Martín-Gayo et al. unpublished results). Notch receptors engage five mammalian surface ligands of the Delta-like (Dll-1, -3 and -4) and Jagged (Jag1 and 2) families expressed on neighbouring cells (Artavanis-Tsakonas et al. 1999; Bray 2006). In the thymus, all Notch ligands except Dll3 are expressed at different thymic niches. Amongst these ligands, Dll4 is specifically expressed on TECs located in the subcapsular and outer cortical regions, while Dll1 expression seems to be restricted to thymic blood vessels (Mohtashami et al. 2010; Hozumi et al. 2008; Koch et al. 2008 and García-León et al. unpublished results). Both Jag1 and Jag2 are expressed in the thymic medulla, and Jag2 is also found in the inner cortex (Heinzel et al. 2007; Mohtashami et al. 2010; Van de Walle et al. 2011; García-León et al. unpublished results).

Notch-ligand binding induces sequential conformational changes in the Notch receptor that expose specific sites for two consecutive cleavages mediated first by ADAM (a disintegrin and metalloprotease) members, ADAM10 (Kuzbanian) and ADAM17 (TNF-alpha converting enzyme or TACE), and then by a γ-secretase protein complex (presenilin, APH-1, nicastrin and PEN-2). As a consequence, the intracellular Notch domain (ICN or NICD) migrates to the nucleus where it associates with the transcription factor CSL (CBF1/RBPJk, Su(H) and LAG-1), displaces co-repressors and recruits co-activators including p300/CBP as well as MAML-1–3, thereafter initiating the transcription of target genes (Kopan and Ilagan 2009). Canonical Notch targets are basic-helix-loop-helix proteins (bHLH) such as Hey and Hes (Jarriault et al. 1995). Other Notch targets include CD25 (Adler et al. 2003), pTα (Reizis and Leder 2002) and molecules involved in cell proliferation and survival such as c-myc (Palomero et al. 2006; Weng et al. 2006), cyclin D1 (Ronchini and Capobianco 2001) and p21/Waf (Rangarajan et al. 2001). More recently, IL-7Rα has been identified as a specific Notch1 target in T-lineage cells (González-García et al. 2009).

The capacity of different Notch ligands and receptors to induce T-cell development has been best illustrated by establishment of a co-culture system that uses a stromal cell line (usually OP9 or S17) made to express individual Notch ligands (Jaleco et al. 2001; Schmitt and Zúñiga-Pflücker 2002), which has been particularly useful for human studies. In this assay, Dll4 appears to be the more effective inducer of T-cell differentiation (Mohtashami et al. 2010) and Dll1 but not Jagged1 can also induce T-cell development from haematopoietic progenitors (Jaleco et al. 2001; Lehar et al. 2005). In contrast, Jag2 appears to have functional similarities to Dll ligands, at least in humans (Van de Walle et al. 2011). These data indicate that different Notch ligands transmit distinct activation signals to T-cell precursors that differentially affect their proliferation and/or differentiation potential, with Jagged ligands (mainly Jagged1) inducing lower Notch activation than Dll ligands (reviewed in Thompson and Zúñiga-Pflücker 2011). Supporting this idea, quantitative Notch signals have been shown to influence the TCR$\alpha\beta$ versus TCR$\gamma\delta$ decision, likely in combination with TCR signals. TCR$\gamma\delta$ development in the mouse seems to be less Notch dependent than TCR$\alpha\beta$ differentiation (Washburn et al. 1997; Ciofani et al. 2006; Garbe et al. 2006), a finding that concurs with a higher dependency of the former on Jagged ligands (Jiang et al. 1998). However, sustained Notch1 signalling in humans has been shown to favour TCR$\gamma\delta$ development at the expense of TCR$\alpha\beta$ generation in vitro (García-Peydró et al. 2003; Van de Walle et al. 2009), although not in vivo in a xenotransplantion assay (García-Peydró et al. unpublished results). These contradictory results underscore the fact that reduction of Notch activation at specific checkpoints is necessary to induce $\alpha\beta$ T-cell differentiation at the expense of $\gamma\delta$ T cells (Van de Walle et al. 2009). Likewise, they emphasise the potential significance of differentially expressed ligands in microenvironmental niches in thymopoiesis. However, only Dll4 has been shown to provide a relevant T-lineage-inducing Notch signal in vivo (Hozumi et al. 2004, 2008; Koch et al. 2008), while the physiological role played by Dll1 and Jagged ligands in the thymus is still unclear. The possibility that important determinants of cellular outcome, such as Notch ligand density and binding affinity could determine an unknown role for these ligands at specific intrathymic locations is particularly appealing (D'Souza et al. 2008), especially because Dll binding affinity can be modulated by glycosylation mediated by the Fringe family of glycosyltransferases expressed in the thymus (Visan et al. 2006; Stanley and Guidos 2009).

3.1 Notch1 Activation Diverts Early Thymic Progenitors from Alternative Cell Fates and Promotes Cell Growth

The critical role of Notch1 in T cell development was first described by Radtke and co-workers, by generating inducible Notch1 loss-of-function mice (Radtke et al. 1999). Total thymocyte numbers were reduced in those mice and T-cell development was blocked at the most immature T-cell stage, while B cells accumulated in the thymus. Complementary to these findings, constitutive expression of active Notch1 (ICN1) in a transplantation setting impaired B-cell differentiation, induced ectopic development of DP thymocytes and subsequently leukaemia (T-ALL) (Pui et al. 1999; Pear et al. 1996). The obvious conclusion of these studies was that Notch1 signalling critically influences the B versus T lineage choice of CLPs. However, as discussed above, LMPP progenitors rather than CLPs include the canonical T-cell precursors in the postnatal thymus, and thus the role that Notch1 signalling plays in early thymopoiesis has been re-examined.

Earlier gain-of-function approaches in humans assessed the impact on ETP multi-lineage potential of either ligand-independent Notch1 signalling, induced by ectopic expression of active ICN1, or ligand-dependent Notch1 activation, supported by OP9-Dll1 stromal cells (Fig. 4). Both strategies provided evidence that the most prominent function of Notch1 signalling is to inhibit non-T cell fates while supporting the expansion of T-lineage progenitors (De Smedt et al. 2002, 2005; García-Peydró et al. 2003, 2006). In the human thymus, these events critically happen at the lymphoid-myeloid branching point and result in impaired generation of myeloid-primed intermediate progenitors (Fig. 2). Consequently, reduced numbers of macrophages and conventional as well as plasmacytoid DCs are generated from ETPs (De Smedt et al. 2005; García-Peydró et al. 2006; Dontje et al. 2006; Martín Gayo et al. unpublished results). In addition, Notch1 signalling blocks the development of NK cells from T/NK lymphoid progenitors, further enforcing T-cell specification (De Smedt et al. 2005; García-Peydró et al. 2006). Complementary loss-of-function approaches using γ-secretase inhibitors (GSI) confirmed these results and showed that increasing thresholds of Notch signalling sequentially suppress B, myeloid/DC and NK cell lineage fates in human thymopoiesis (De Smedt et al. 2005).

Notch-mediated suppression of myeloid cell fate in human lympho-myeloid progenitors is actively induced by repression of myeloid gene transcription that, importantly, seems to be ligand specific (Martín-Gayo et al. unpublished results). In addition, Notch1 signalling triggers a T-cell lineage gene program in humans (García-Peydró et al. 2006; Van de Walle et al. 2011), coincident with the profile reported in mice (reviewed in Rothenberg et al. 2008; Thompson and Zúñiga-Pflücker 2011), a finding that supports an instructive role of Notch1 activation in T-cell specification. Notably, very recent results have shown that such an instructive role crucially relies on a transcription factor, TCF1, induced by Notch signals. TCF1 imposes the T-cell fate by up-regulating expression of other transcription factors essential for T-cell differentiation such as Gata3 and Bcl11b, and

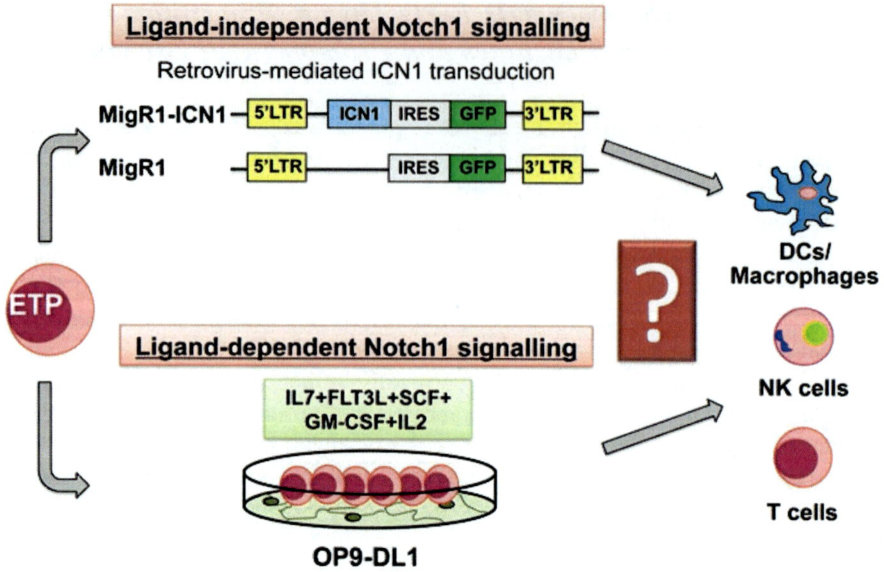

Fig. 4 *Experimental strategies used to analyse the impact of Notch1 signalling in human T-cell development.* Ligand-independent constitutive Notch signalling is induced in human ETPs by retroviral transduction using a bicistronic vector encoding active Notch1 (ICN1) and green-fluorescent protein (GFP) as a cell tracer. For ligand-dependent Notch activation, ETPs are co-cultured onto OP9 stromal cells ectopically expressing the Dll1 Notch ligand (OP9–DL1). In both systems, cells are cultured in the presence of a multilineage-cytokine cocktail that supports the simultaneous differentiation of lymphoid (T and NK cells) and myeloid (macrophages and DCs) lineages from ETPs

TCR components (Weber et al. 2011). However, ICN1 gene targets that control T-cell proliferation are not similarly triggered by TCF1, and Notch1-induced proliferation may, thus, be TCF-1-independent. Indeed, proliferation induced by Notch1 was shown to rely on unique signals provided by cytokines (Varnum-Finney et al. 2003), specifically by IL-7 (García-Peydró et al. 2006; González-García et al. 2009). Accordingly, Notch1-induced proliferation correlates with up-regulation of IL-7R expression (García-Peydró et al. 2006; González-García et al. 2009). As a whole, these data indicate that acquisition of a functional IL-7R marks the critical checkpoint of inhibition of non-T cell potentials and T-lineage specification of lympho-myeloid progenitors induced by Notch, and also controls the pivotal stage of cellular expansion of T-cell committed progenitors (de Yébenes et al. 2002; García-Peydró et al. 2006).

Once T-cell specification is induced, definitive T-cell differentiation and growth functions triggered downstream of Notch1 become uncoupled events. It has been shown that enforced expression of active Notch1 in human ETPs is itself sufficient to block non-T cell development and to trigger an almost unlimited cellular expansion in vitro in response to IL-7; however, it is insufficient to induce TCR rearrangements and to complete T-cell maturation of DN2 thymocytes, unless cocultured with stromal cells (García-Peydró et al. 2006). Likewise, a DN2 developmental arrest has recently been observed when murine fetal liver progenitors are activated with immobilised Dll4 plus cytokines under stroma-free conditions (Ikawa et al. 2010). These results suggest a requirement of additional inductive signals provided by stromal cells supporting a complete differentiation program along the T-cell lineage. Alternatively, Notch1 signalling could provide self-renewal capacity to DN2 cells, and simultaneously arrest development beyond the DN2 stage, thus allowing for segregation of proliferation and differentiation processes during thymopoiesis. Supporting the latter possibility, an early decrease in Notch activation is required for human thymic precursors to complete $\alpha\beta$ T-lineage differentiation in vitro (Van de Walle et al. 2009). Moreover, in vitro production of mature $\alpha\beta$-lineage cells from murine thymic precursors is contingent to decreased IL-7 responsiveness and self-renewal arrest beyond the DN2 stage (Balciunaite et al. 2005). Diminished IL-7R signalling was recently shown to be necessary for DN2 mouse thymocytes to up-regulate Bcl11b, a transcription factor that is essential to drive full development along the T-cell lineage (Ikawa et al. 2010; Li et al. 2010). Therefore, it is likely that physiological progression through the DN2-determination step is instructed by environmental signals in the thymus, such as limited IL-7 availability and/or reduced IL-7R expression. According to this view, Bcl11b deficiency selectively impairs development of $\alpha\beta$ but not $\gamma\delta$ T cells (Ikawa et al. 2010), which develop from DN2 thymocytes expressing either low or high IL-7R levels, respectively (Kang et al. 2001). Collectively, these data suggest that regulation of Bcl11b expression is an early T-cell developmental checkpoint controlled by IL-7R-mediated signalling.

3.2 *Notch1 Signalling Controls T-Lineage-Specific IL-7R Expression in Early Thymopoiesis*

The stage- and lineage-specific role of IL-7 during thymopoiesis indicates that strict mechanisms control the dynamic intrathymic regulation of IL-7R expression. Likewise, regulatory mechanisms may control the differential expression of IL-7R in T- and B-lineage cells during lymphopoiesis. In mouse early lymphoid/B-cell progenitors, IL-7Rα gene (*Il7ra*) transcription is specifically regulated by the Ets transcription factor PU.1 (DeKoter et al. 2002). PU.1 is expressed very early in thymopoiesis as well, but PU.1 down-regulation is obligatory for T-cell fate specification and progression along the T-cell lineage (Anderson et al. 2002). PU.1 function in mature T-cells seems to be replaced by the Ets transcription factor GA binding protein (GABP), but its role during early thymopoiesis is less clear (Xue et al. 2004). In B lymphopoiesis, however, GABP cooperates with PU.1 and regulates IL-7Rα expression in pre-B and committed B-cells (DeKoter et al. 2007). While these data support the existence of specific regulators of IL-7Rα expression in B-cell development, the nature of equivalent regulators in the T-cell lineage has been an open question for years. Recently, molecular studies from our group have

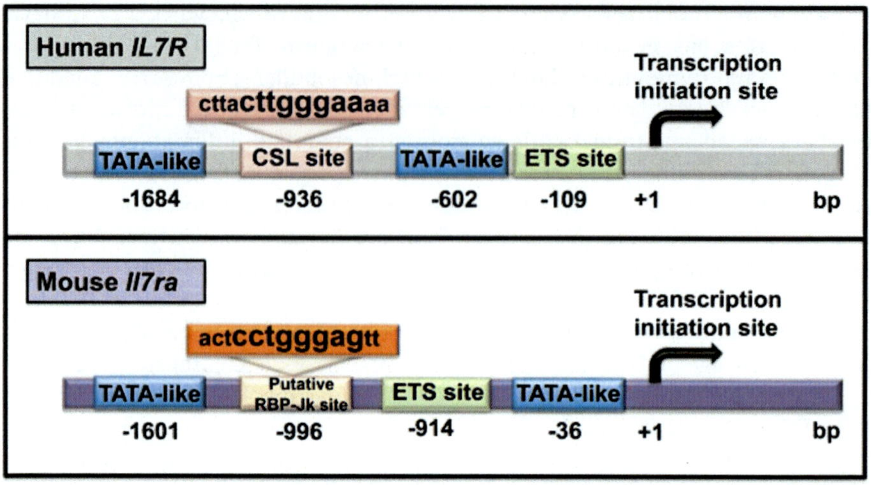

Fig. 5 *Schematic representation of human IL7R and mouse Il7ra gene promoters.* Human *IL7R* and mouse *Il7ra* gene promoters expand about 2,000 base pairs (bp) upstream of the transcription initiation site. Both promoters share two TATA-like boxes and an ETS transcription factor-binding site. The human *IL7R* promoter also contains a CSL-binding site that is conserved in the mouse *IL7ra* promoter. The physiological also relevance of the latter has yet not been confirmed

shown that Notch1 accomplishes this function at least in humans (González-García et al. 2009). Both gain- and loss-of-function approaches have shown that expression of IL-7Rα in developing T-lineage cells critically depends on Notch1 activation. Notch1 regulates IL-7Rα expression at the transcriptional level and, notably, in a T-lineage-specific manner, since IL-7Rα gene (*IL7R*) transcription can be inhibited by ectopic expression of a dominant negative form of the replace by:MAML-1 co-activator (dnMAML-1) in T- but not B-lineage cells (González-García et al. 2009). Chromatine immunoprecipitation and luciferase reporter assays have further established that *IL7R* gene expression is directly induced by active Notch1 in T-cell lines and DN2 thymocytes, indicating that *IL7R* is a transcriptional target of Notch1. Supporting the participation of CSL in Notch1-induced *IL7R* transcription, we have identified a conserved CSL-binding site in the *IL7R* promoter (Fig. 5) and show that either site-directed mutagenesis or CSL-deficiency impairs *IL7R* promoter activity induced by active Notch1. Therefore, T-lineage-specific *IL7R* transcription induced by Notch1 is CSL/MAML-1-dependent (González-García et al. 2009). More recently, ICN1 has been shown to interact with an *IL7R* gene enhancer in a human T-cell line, suggesting that additional mechanisms of regulation of IL-7Rα expression mediated by Notch1 could exist (Wang et al. 2011).

The physiological role of Notch1 in the regulation of IL-7Rα expression in human thymopoiesis is supported by studies showing that, which encodes the IL-7R effector bcl-2, Notch1 activity parallels expression of *IL7R* as well as *BCL2* from the ETP to DN3 stages. Conversely, transcription of *NOTCH1* and *IL7R* decreases concordantly beyond the DN3 stage and remains low throughout the rest of thymocyte development. Accordingly, defective Notch1 activation selectively results in a compromised expansion of the DN1-3 compartments, which can be rescued by ectopic IL-7Rα expression, suggesting that Notch1 signals are no longer required once T-cell specification and IL-7R expression have been induced. However, IL-7R is unable to replace Notch1 signals at the β-selection checkpoint (González-García et al. 2009), a finding that concurs with the reported requirement of Notch1 and pre-TCR signalling during β-selection (Wolfer et al. 2002; Ciofani et al. 2004; Ciofani and Zúñiga-Pflücker 2005; Maillard et al. 2006; Taghon et al. 2009). Therefore, Notch1 signals control thymocyte proliferation at two sequential checkpoints. First, a functional IL-7R is up-regulated on T-cell specified progenitors and second, expression of a pre-TCR complex is induced that supports metabolism, survival and proliferation of committed T-cell progenitors as well as progression to the DP stage independently of IL-7R. Between both proliferation phases, decreased Notch1 signalling seems to be required to induce down-regulation of IL-7R expression and Bcl11b up-regulation, which will finally allow T-cell-specified progenitors to complete differentiation along the $\alpha\beta$ lineage, as discussed above.

4 Notch1 Signalling in T-ALL

T-cell acute lymphoblastic leukaemia (T-ALL) is a lymphoproliferative disorder accounting for 10–15 % of pediatric and 25 % of adult ALL cases, which results from the malignant transformation of normal developing T cells in the thymus (Pui et al. 2004). Aberrant Notch1 signalling was initially described in human T-ALLs (<1 %) with rare chromosomal translocations that generate a truncated Notch1 isoform lacking the extracellular domain (TAN1) under the transcriptional control of the TCRβ enhancer (Ellisen et al. 1991). Subsequently, the group of Aster provided evidence that Notch1 signals play a more prominent role in leukaemogenesis than initially suspected, as activating *NOTCH1* mutations were found in more than 50 % of human T-ALLs (Weng et al. 2004). These mutations involve the extracellular heterodimerisation (HD) domain and, less frequently, the C-terminal PEST domain of Notch1. HD mutations increase ADAM cleavage and subsequent ligand-independent receptor activation, while PEST mutations increase the stability and half-life of ICN1. Other Notch1 activating mutations as well as mutants of genes that regulate turnover of ICN1 such as *Fbw7,* which encodes a ubiquitin ligase involved in ICN1 degradation, have been subsequently identified in mouse models of T-ALL and T-ALL patients (reviewed in Aifantis et al. 2008; Li and von Boehmer 2011; Aster et al. 2011; Koch and Radtke 2011). Thus, there is an increasing interest in understanding the role of Notch1 in the pathogenesis of T-ALL, with the final aim of identifying novel therapies which target Notch1 signalling.

The main targets of aberrant Notch1 activation leading to leukaemia are Notch-associated signalling pathways that control survival and proliferation in normal T-cell development. As highlighted above, the IL-7R and pre-TCR are the major Notch-dependent pathways that accomplish this function in thymopoiesis. Historically, the pre-TCR pathway was suggested first to interact with Notch1 signals in T-cell oncogenesis, as mice transplanted with BM progenitors expressing active *NOTCH1* alleles rapidly developed an aggressive T-cell leukaemia (Pear et al. 1996; Pui et al. 1999), but only when pre-TCR signalling was intact (Allman et al. 2001). Likewise, T-ALL development in Notch3 transgenic mice was shown to be dependent on pre-TCR expression (Bellavia et al. 2002). However, a cooperative rather than absolute requirement of pre-TCR in Notch-induced leukaemogenesis has been proposed later (Campese et al. 2006). Regarding possible molecular mechanisms underlying this cooperation, it was found that the *PTCRA* gene encoding pTα is a transcriptional target of Notch (Reizis and Leder 2002). In addition, other pre-TCR components including CD3ε and TCRβ seem to be regulated by Notch signals as well, further suggesting that the Notch pathway could be upstream of pre-TCR assembly and expression (Aifantis et al. 2008). Nonetheless, both pathways activate common transcription factors and kinases, and share BCL-2A1 and cyclin D3 as common targets involved in G1/S cell cycle progression. This would suggest that they can act in parallel but converge at signalling intermediates in T-ALLs (Aifantis et al. 2008). Supporting this possibility, c-myc, a crucial regulator of cellular metabolism and cell cycle progression, whose expression picks around the β-selection checkpoint, has been identified as a key Notch target that cooperates with ICN1 in Notch1-dependent leukaemogenesis (Weng et al. 2006; Palomero et al. 2006). Importantly, Notch1 and c-myc activate common targets required for growth of leukaemic cells, suggesting a feed-forward loop in leukaemogenesis (Palomero et al. 2006). Notch activation positively regulates activity of the mTOR pathway in a c-myc-dependent manner (Chan et al. 2007). The PTEN/PI3K/Akt/mTOR pathway is a major pre-TCR-associated pro-oncogenic pathway regulated by Notch1 in T-ALL. Seminal work by the group of Ferrando showed that Notch1 is a negative regulator of PTEN, and identified recurrent *PTEN* inactivating mutations in T-ALLs that conferred GSI resistance (Palomero et al. 2007). The consequence of PTEN loss is a deregulated balance among activation/inhibition of PI3K and aberrant activation of Akt, a major inducer of proliferation and survival in T cells. In addition, *Pi3K* or *Akt* mutations have been identified in a high proportion of T-ALLs. Notch1 can also induce PI3K/Akt-dependent proliferation by inhibition of p53, another tumor suppressor downstream of pre-TCR signaling (Mungamuri et al. 2006). Other key downstream effectors of pre-TCR signalling activated by Notch1 in T-ALL include the NF-κB and NFAT pathways (Vilimas et al. 2007; Ciofani and Zúñiga-Pflücker 2005; Aifantis et al. 2001). Finally, it has been suggested that Notch promotes inhibition of the transcriptional activity of the E2A proteins in T-ALL, through a mechanism that involves pre-TCR-mediated ERK-dependent up-regulation of the E2A inhibitors Id1 or Id3, while pre-TCR-mediated induction of Id3 represses E2A-dependent transcription of *Notch1* under physiological conditions (reviewed in Li and von Boehmer 2011). Therefore, multiple signalling pathways driven by Notch1 and pre-TCR interact synergistically to promote transformation to T-ALL (reviewed in Koch and Radtke 2011; Li and von Boehmer 2011). These data highlight the requirement of Notch1 down-regulation after β-selection to avoid the oncogenic properties of Notch signalling.

4.1 Notch1 and IL-7R: Independent or Complementary Pathways in T-ALL?

As discussed above, another pathway critically involved in physiological growth of developing thymocytes is the IL-7R signalling pathway. Several results have suggested that IL-7 and IL-7R may contribute to T-cell leukaemia progression. In mice, expression of an IL-7 transgene results in lymphoma development (Rich et al. 1993), and AKR/J mice that show an up-regulated expression of IL-7Rα in the thymus develop spontaneous thymomas (Laouar et al. 2004). More importantly, human T-ALLs commonly express functional IL-7R that significantly contributes to T-ALL proliferation in response to exogenous IL-7 (Dibirdik et al. 1991; Barata et al. 2005). Notably, PI3K is a major effector of IL-7-induced viability and proliferation of T-ALLs (Barata et al. 2004), a finding that places the PI3K/Akt pathway at the crossroads of Notch1 and IL-7R signalling in T-ALL. Other effectors of IL-7R signalling including Bcl2 and cyclin D1 are also over-expressed in T-ALL (Barata et al. 2005). The recent identification of *IL7R* as a downstream target of Notch1 activity in normal human T-cell development might suggest a functional link between the Notch1 and IL-7R pathways in T-cell leukaemogenesis. Confirming this possibility, we found that *IL7R* is transcriptionally regulated by Notch1 activity also in T-ALL. IL-7Rα expression is, specifically, down-regulated in T-ALL but not B-ALL cell lines when Notch1 signalling is inhibited by GSI treatment or ectopic expression of dnMAML-1 (González-García et al. 2009). Notch inhibition results in decreased proliferation and cell cycle arrest, likely involving PTEN up-regulation (Palomero et al. 2007). Significantly, impaired proliferation of these T-ALL cell lines can be rescued by ectopic expression of IL-7Rα, which results in a selective growth advantage of IL-7Rα-expressing T-ALL cells in response to IL-7 (González-García et al. 2009). Therefore, IL-7/IL-7R signalling is able to support the survival and expansion of leukaemic cells with impaired Notch1 signalling. More importantly, our recent studies indicate that IL-7R is an important mediator of cell growth in primary T-ALLs as well (González-García et al. unpublished results).

Overall, these results support a cooperative role of Notch1 and IL-7R pathways in supporting leukaemogenesis. Likewise, they point to the IL-7/IL-7R pathway as a potential candidate to induce and/or maintain T-cell leukaemogenesis independently of Notch signals. The latter possibility is further supported by the observation that 18 % of adult and 2 % of pediatric T-ALL cases have activating mutations in *JAK1* (Flex et al. 2008), which encodes a tyrosine kinase that directly binds IL-7R and promotes signalling (Fig. 3). Therefore, it can be hypothesised

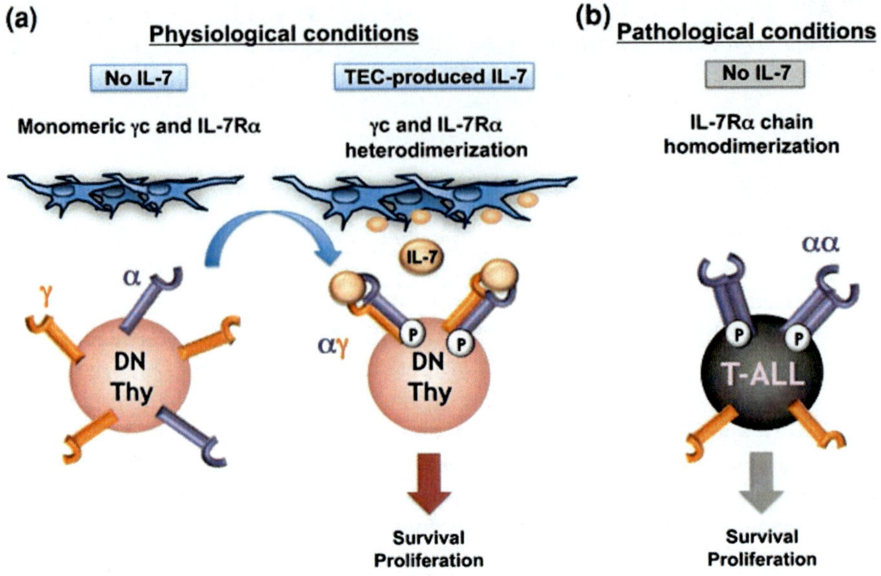

Fig. 6 *Mechanisms of IL-7R activation under physiological and pathological conditions.* **a** Under physiological conditions, double negative thymocytes (DN Thy) express both IL-7Rα and γc chains in monomeric conformation. Binding of IL-7 produced by TECs to the IL-7R components promotes their heterodimerisation, and thereby triggers IL-7R signalling leading to survival and proliferation. **b** In pathological conditions, as in T-ALL, *IL7R* mutations at the juxtamembrane domain result in the formation of stable IL-7Rα-IL-7Rα homodimers, which promote signalling in the absence of IL-7, thus inducing uncontrolled survival and proliferation

that *IL7R* itself might be a target of activating mutations in T-ALLs. Based on analyses of the complete coding sequence of *IL7R* in pediatric T-ALL samples, others and we have recently provided evidence that heterozygous oncogenic gain-of-function mutations do, in fact, occur in around 10 % of T-ALLs (Zenatti et al. 2011; Shochat et al. 2011). Notably, *IL7R* mutations do not occur in the cytoplasmic tail that recruits signalling effectors, but consist of in-frame insertions or deletions insertions in the juxtamembrane-transmembrane domain at the interface with the extracellular region. The vast majority of *IL7R* mutations create an unpaired cysteine residue that results in disulfide-bond-mediated homo-dimerisation of IL-7Rα chains able to signal in the absence of γc and IL-7 ligand binding (Fig. 6). Interestingly, *IL7R* mutations were also found by Shochat and coworkers in B-ALL samples, suggesting a general strategy for mutational activation of type I cytokine receptors in leukaemia. Therefore, although therapeutic strategies directed to Notch1 inhibition, particularly treatment with GSI, initially emerged as a promising therapy (Weng et al. 2004), recent knowledge of the molecular pathology of T-ALL open new avenues for the design of specific targeted therapies.

5 Conclusions

The study of the molecular mechanisms underlying T-cell development and transformation highlights the close relationship between Notch signalling pathways involved in T-cell physiology and pathology. The fundamental function of Notch1 in thymopoiesis is to drive T-cell specification from multipotent precursors seeding the thymus and to support further progression along the T-cell lineage. These processes are associated with a unique role of Notch1 as a crucial regulator of cellular expansion at two critical checkpoints. Firstly, Notch1 controls IL-7R-dependent expansion of the T-cell-specified progenitor pool, and thereafter Notch1 cooperates with the pre-TCR to trigger expansion of progenitors that successfully progress along the T-cell maturation pathway. However, deregulated Notch1 activation at these stages results in T-cell transformation and leukaemia. Therefore, the identification of Notch1 signalling effectors involved in physiological proliferation of developing T-cells is crucial for designing new therapeutic strategies that target relevant oncogenic pathways in T-ALL. The identification of the IL-7R as a downstream transcriptional target of Notch1 both in physiology and pathology, together with the finding of oncogenic gain-of-function *IL-7R* mutations in T-ALL, open new possibilities for the development of specific targeted therapies. Ongoing preclinical studies exploring the efficacy of new therapies targeting the IL-7R in vivo in xenograft models of human T-ALL offer promise for the development of more effective T-ALL treatments in the near future.

Acknowledgments This work was supported by grants from Plan Nacional (SAF2010-15106), Subprograma de Fomento de la Cooperación Científica Internacional (PLE-2009-0110), Fundación Sandra Ibarra and Instituto de Salud Carlos III (RD06/0014/1012) and by an institutional grant from the Fundación Ramón Areces. S.G–G was supported by Ministerio de Ciencia e Innovación (MICINN) (FPI program). The authors declare no conflicting financial interests.

References

Adler SH, Chiffoleau E, Xu L, Dalton NM, Burg JM, Wells AD, Wolfe MS, Turka LA, Pear WS (2003) Notch signaling augments T cell responsiveness by enhancing CD25 expression. J Immunol 171:2896–2903

Adolfsson J, Månsson R, Buza-Vidas N, Hultquist A, Liuba K, Jensen CT, Bryder D, Yang L, Borge OJ, Thoren LA, Anderson K, Sitnicka E, Sasaki Y, Sigvardsson M, Jacobsen SE (2005) Identification of Flt3 + lympho-myeloid stem cells lacking erythro-megakaryocytic potential a revised road map for adult blood lineage commitment. Cell 121:295–306

Aifantis I, Gounari F, Scorrano L, Borowski C, von Boehmer H (2001) Constitutive pre-TCR signaling promotes differentiation through Ca2 + mobilization and activation of NF-kappaB and NFAT. Nat Immunol 2:403–409

Aifantis I, Raetz E, Buonamici S (2008) Molecular pathogenesis of T-cell leukaemia and lymphoma. Nat Rev Immunol 8:380–390

Akashi K, Kondo M, von Freeden-Jeffry U, Murray R, Weissman IL (1997) Bcl-2 rescues T lymphopoiesis in interleukin-7 receptor-deficient mice. Cell 89:1033–1034

Akashi K, Kondo M, Weissman IL (1998) Role of interleukin-7 in T-cell development from hematopoietic stem cells. Immunol Rev 165:13–28
Akashi K, Richie LI, Miyamoto T, Carr WH, Weissman IL (2000) B lymphopoiesis in the thymus. J Immunol 164:5221–5226
Allman D, Karnell FG, Punt JA, Bakkour S, Xu L, Myung P, Koretzky GA, Pui JC, Aster JC, Pear WS (2001) Separation of Notch 1 promoted lineage commitment and expansion/transformation in developing T cells. J Exp Med 194:99–106
Allman D, Sambandam A, Kim S, Miller JP, Pagan A, Well D, Meraz A, Bhandoola A (2003) Thymopoiesis independent of common lymphoid progenitors. Nat Immunol 4:168–174
Anderson G, Lane PJ, Jenkinson EJ (2007) Generating intrathymic microenvironments to establish T-cell tolerance. Nat Rev Immunol 7:954–963
Anderson MK, Weiss AH, Hernandez-Hoyos G, Dionne CJ, Rothenberg EV (2002) Constitutive expression of PU.1 in fetal hematopoietic progenitors blocks T cell development at the pro-T cell stage. Immunity 16:285–296
Ardavin C, Wu L, Li CL, Shortman K (1993) Thymic dendritic cells and T cells develop simultaneously in the thymus from a common precursor population. Nature 362:761–763
Artavanis-Tsakonas S, Rand MD, Lake RJ (1999) Notch signaling: cell fate control and signal integration in development. Science 284:770–776
Aster JC, Blacklow SC, Pear WS (2011) Notch signalling in T-cell lymphoblastic leukemia/lymphoma and other haematological malignancies. J Patol 223:262–273
Balciunaite G, Ceredig R, Fehling HJ, Zúñiga-Pflücker JC, Rolink AG (2005) The role of Notch and IL-7 signaling in early thymocyte proliferation and differentiation. Eur J Immunol 35:1292–1300
Barata JT, Silva A, Brandao JG, Nadler LM, Cardoso AA, Boussiotis VA (2004) Activation of PI3K is indispensable for interleukin 7-mediated viability, proliferation, glucose use, and growth of T cell acute lymphoblastic leukemia cells. J Exp Med 200:659–669
Barata JT, Cardoso AA, Boussiotis VA (2005) Interleukin-7 in T-cell acute lymphoblastic leukemia: an extrinsic factor supporting leukemogenesis? Leuk Lymphoma 46:483–495
Bell JJ, Bhandoola A (2008) The earliest thymic progenitors for T cells possess myeloid lineage potencial. Nature 452:764–767
Bellavia D, Campese AF, Checquolo S, Balestri A, Biondi A, Cazzaniga G, Lendahl U, Fehling HJ, Hayday AC, Frati L, von Boehmer H, Gulino A, Screpanti I (2002) Combined expression of pTalpha and Notch3 in T cell leukemia identifies the requirement of preTCR for leukemogenesis. Proc Natl Acad Sci U S A 99:3788–3793
Benz C, Bleul CC (2005) A multipotent precursor in the thymus maps to the branching point of the T versus B lineage decision. J Exp Med 202:21–31
Benz C, Martins VC, Radtke F, Bleul CC (2008) The stream of precursors that colonizes the thymus proceeds selectively through the early T lineage precursor stage of T cell development. J Exp Med 205:1187–1199
Bhandoola A, von Boehmer H, Petrie HT, Zúñiga-Pflücker JC (2007) Commitment and developmental potential of extrathymic and intrathymic T cell precursors: plenty to choose from. Immunity 26:678–689
Blom B, Spits H (2006) Development of human lymphoid cells. Annu Rev Immunol 24:287–320
Blom B, Verschuren MC, Heemskerk MH, Bakker AQ, van Gastel-Mol EJ, Wolvers-Tettero IL, van Dongen JJ, Spits H (1999) TCR gene rearrangements and expression of the pre-T cell receptor complex during human T-cell differentiation. Blood 93:3033–3043
Bray SJ (2006) Notch signalling: a simple pathway become complex. Nat Rev Mol Cell Biol 7:678–689
Campese AF, Garbe AI, Zhang F, Grassi F, Screpanti I, von Boehmer H (2006) Notch1-dependent lymphomagenesis is assisted by but does not essentially require pre-TCR signaling. Blood 108:305–310
Carrasco YR, Trigueros C, Ramiro AR, de Yébenes VG, Toribio ML (1999) Beta-selection is associated with the onset of CD8beta chain expression on CD4(+)CD8alphaalpha(+) pre-T cells during human intrathymic development. Blood 94:3491–3498

Carrasco YR, Ramiro AR, Trigueros C, de Yébenes VG, García-Peydró M, Toribio ML (2001) An endoplasmic reticulum retention function for the cytoplasmic tail of the human pre-T cell receptor (TCR) alpha chain: potential role in the regulation of cell surface pre-TCR expression levels. J Exp Med 193:1045–1058

Carrasco YR, Navarro MN, de Yébenes VG, Ramiro AR, Toribio ML (2002) Regulation of surface expression of the human pre-T cell receptor complex. Semin Immunol 14:325–334

Ceredig R (2012) Fates and potentials of thymus-seeding progenitors. Nat Immunol 13:309–310

Chan SM, Weng AP, Tibshirani R, Aster JC, Utz PJ (2007) Notch signals positively regulate activity of the mTOR pathway in T-cell acute lymphoblastic leukemia. Blood 110:278–286

Ciofani M, Zúñiga-Pflücker JC (2005) Notch promotes survival of pre-T cells at the beta-selection checkpoint by regulating cellular metabolism. Nat Immunol 6:881–888

Ciofani M, Zúñiga-Pflücker JC (2007) The thymus as an inductive site for T lymphopoiesis. Annu Rev Cell Dev Biol 23:463–493

Ciofani M, Schmitt TM, Ciofani A, Michie AM, Cuburu N, Aublin A, Maryanski JL, Zúñiga-Pflücker JC (2004) Obligatory role for cooperative signaling by pre-TCR and Notch during thymocyte differentiation. J Immunol 172:5230–5239

Ciofani M, Knowles GC, Wiest DL, von Boehmer H, Zuniga-Pflucker JC (2006) Stage-specific and differential notch dependency at the $\alpha\beta$ and $\gamma\delta$ T lineage bifurcation. Immunity 25:105–116

De Smedt M, Reynvoet K, Kerre T, Taghon T, Verhasselt B, Vandekerckhove B, Leclercq G, Plum J (2002) Active form of Notch imposes T cell fate in human progenitor cells. J Immunol 169:3021–3029

De Smedt M, Hoebeke I, Reynvoet K, Leclercq G, Plum J (2005) Different thresholds of Notch signaling bias human precursor cells toward B-, NK-, monocytic/dendritic-, or T-cell lineage in thymus microenvironment. Blood 106:3498–3506

de Yébenes VG, Carrasco YR, Ramiro AR, Toribio ML (2002) Identification of a myeloid intrathymic pathway of dendritic cell development marked by expression of the granulocyte macrophage-colony-stimulating factor receptor. Blood 99:2948–2956

DeKoter RP, Lee HJ, Singh H (2002) PU.1 regulates expression of the interleukin-7 receptor in lymphoid progenitors. Immunity 16:297–309

DeKoter RP, Schweitzer BL, Kamath MB, Jones D, Tagoh H, Bonifer C, Hildeman DA, Huang KJ (2007) Regulation of the interleukin-7 receptor alpha promoter by the Ets transcription factors PU.1 and GA-binding protein in developing B cells. J Biol Chem 282:14194–14204

Dibirdik I, Langlie MC, Ledbetter JA, Tuel-Ahlgren L, Obuz V, Waddick KG, Gajl-Peczalska K, Schieven GL, Uckun FM (1991) Engagement of interleukin-7 receptor stimulates tyrosine phosphorylation, phosphoinositide turnover, and clonal proliferation of human T-lineage acute lymphoblastic leukemia cells. Blood 78:564–570

Dontje W, Schotte R, Cupedo T, Nagasawa M, Scheeren F, Gimeno R, Spits H, Blom B (2006) Delta-like1-induced Notch1 signaling regulates the human plasmacytoid dendritic cell versus T-cell lineage decision through control of GATA-3 and Spi-B. Blood 107:2446–2452

Doulatov S, Notta F, Eppert K, Nguyen LT, Ohashi PS, Dick JE (2010) Revised map of the human progenitor hierarchy shows the origin of macrophages and dendritic cells in early lymphoid development. Nat Immunol 11:585–593

D'Souza B, Miyamoto A, Weinmaster G (2008) The many facets of Notch ligands. Oncogene 27:5148–5167

Durum SK, Candeias S, Nakajima H, Leonard WJ, Baird AM, Berg LJ, Muegge K (1998) Interleukin 7 receptor control of T cell receptor γ gene rearrangement: role of receptor-associated chains and locus accessibility. J Exp Med 188:2233–2241

Ellisen LW, Bird J, West DC, Soreng AL, Reynolds TC, Smith SD, Sklar J (1991) TAN-1, the human homolog of the Drosophila notch gene, is broken by chromosomal translocations in T lymphoblastic neoplasms. Cell 66:649–661

Fiorini E, Merck E, Wilson A, Ferrero I, Jiang W, Koch U, Auderset F, Laurenti E, Tacchini-Cottier F, Pierres M, Radtke F, Luther SA, Macdonald HR (2009) Dynamic regulation of Notch 1 and Notch 2 surface expression during T cell development and activation revealed by novel monoclonal antibodies. J Immunol 183:7212–7222

Flex E, Petrangeli V, Stella L, Chiaretti S, Hornakova T, Knoops L, Ariola C, Fodale V, Clappier E, Paoloni F, Martinelli S, Fragale A, Sanchez M, Tavolaro S, Messina M, Cazzaniga G, Camera A, Pizzolo G, Tornesello A, Vignetti M, Battistini A, Cavé H, Gelb BD, Renauld JC, Biondi A, Constantinescu SN, Foà R, Tartaglia M (2008) Somatically acquired JAK1 mutations in adult acute lymphoblastic leukemia. J Exp Med 205:751–758

Fowlkes BJ, Edison L, Mathieson BJ, Chused TM (1985) Early T lymphocytes. Differentiation in vivo of adult intrathymic precursor cells. J Exp Med 162:802–822

Galy A, Verma S, Bárcena A, Spits H (1993) Precursors of CD3+CD4+CD8+ cells in the human thymus are defined by expression of CD34. Delineation of early events in human thymic development. J Exp Med 178:391–401

Garbe AI, Krueger A, Gounari F, Zúñiga-Pflücker JC, von Boehmer H (2006) Differential synergy of Notch and T cell receptor signaling determines alphabeta versus gammadelta lineage fate. J Exp Med 203:1579–1590

García-Peydró M, de Yébenes VG, Toribio ML (2003) Sustained Notch1 signaling instructs the earliest human intrathymic precursors to adopt a gammadelta T-cell fate in fetal thymus organ culture. Blood 102:2444–2451

García-Peydró M, de Yébenes VG, Toribio ML (2006) Notch1 and IL-7 receptor interplay maintains proliferation of human thymic progenitors while suppressing non-T cell fates. J Immunol 177:3711–3720

González-García S, García-Peydró M, Martín-Gayo E, Ballestar E, Esteller M, Bornstein R, de la Pompa JL, Ferrando AA, Toribio ML (2009) CSL-MAML-dependent Notch1 signaling controls T lineage-specific IL-7R {alpha} gene expression in early human thymopoiesis and leukemia. J Exp Med 206:779–791

Gossens K, Naus S, Corbel SY, Lin S, Rossi FM, Kast J, Ziltener HJ (2009) Thymic progenitor homing and lymphocyte homeostasis are linked via S1P-controlled expression of thymic P-selectin/CCL25. J Exp Med 206:761–778

Hao QL, George AA, Zhu J, Barsky L, Zielinska E, Wang X, Price M, Ge S, Crooks GM (2008) Human intrathymic lineage commitment is marked by differential CD7 expression: identification of CD7- lympho-myeloid thymic progenitors. Blood 111:1318–1326

Harman BC, Jenkinson WE, Parnell SM, Rossi SW, Jenkinson EJ, Anderson G (2005) T/B lineage choice occurs prior to intrathymic Notch signaling. Blood 106:886–892

Heinzel K, Benz C, Martins VC, Haidl ID, Bleul CC (2007) Bone marrow-derived hemopoietic precursors commit to the T cell lineage only after arrival in the thymic microenvironment. J Immunol 178:858–868

Hozumi K, Negishi N, Suzuki D, Abe N, Sotomaru Y, Tamaoki N, Mailhos C, Ish-Horowicz D, Habu S, Owen MJ (2004) Delta-like 1 is necessary for the generation of marginal zone B cells but not T cells in vivo. Nat Immunol 5:638–644

Hozumi K, Maihlos G, Negishi N, Hirano K, Yahata T, Ando K, Zuklys S, Holländer GA, Shima DT, Habu S (2008) Delta-like 4 is indispensable in thymic environment specific for T cell development. J Exp Med 205:2507–2513

Ikawa T, Hirose S, Masuda K, Kakugawa K, Satoh R, Shibano-Satoh A, Kominami R, Katsura Y, Kawamoto H (2010) An essential developmental checkpoint for production of the T cell lineage. Science 329:93–96

Jaleco AC, Neves H, Hooijberg E, Gameiro P, Clode N, Haury M, Henrique D, Parreira L (2001) Differential effects of Notch ligands Delta-1 and Jagged-1 in human lymphoid differentiation. J Exp Med 194:991–1002

Jarriault S, Brou C, Logeat F, Schroeter EH, Kopan R, Israel A (1995) Signalling downstream of activated mammalian Notch. Nature 377:355–358

Jiang R, Lan Y, Chapman HD, Shawber C, Norton CR, Serreze DV, Weinmaster G, Gridley T (1998) Defects in limb, craniofacial, and thymic development in Jagged2 mutant mice. Genes Dev 12:1046–1057

Kang J, Volkmann A, Raulet DH (2001) Evidence that gammadelta versus alphabeta T cell fate determination is initiated independently of T cell receptor signaling. J Exp Med 193:689–698

Kawamoto H, Katsura Y (2009) A new paradigm for hematopoietic cell lineages: revision of the classical concept of the myeloid-lymphoid dichotomy. Trends Immunol 30:193–200

Koch U, Radtke F (2011) Mechanisms of T cell development and transformation. Annu Rev Cell Dev Biol 10:539–562

Koch U, Fiorini E, Benedito R, Besseyrias V, Schuster-Gossler K, Pierres M, Manley NR, Duarte A, Macdonald HR, Radtke F (2008) Delta-like 4 is the essential, nonredundant ligand for Notch1 during thymic T cell lineage commitment. J Exp Med 205:2515–2523

Kondo M, Weissman IL, Akashi K (1997) Identification of clonogenic common lymphoid progenitors in mouse bone marrow. Cell 91:661–672

Kopan R, Ilagan MX (2009) The canonical Notch signaling pathway: unfolding the activation mechanism. Cell 137:216–233

Kraft DL, Weissman IL, Waller EK (1993) Differentiation of CD3-4-8- human fetal thymocytes in vivo: characterization of a CD3-4+8- intermediate. J Exp Med 178:265–277

Kurtzberg J, Denning SM, Nycum LM, Singer KH, Haynes BF (1989) Immature human thymocytes can be driven to differentiate into nonlymphoid lineages by cytokines from thymic epithelial cells. Proc Natl Acad Sci U S A 86:7575–7579

Laouar Y, Crispe IN, Flavell RA (2004) Overexpression of IL-7Rα provides a competitive advantage during early T-cell development. Blood 103:1985–1994

Lehar SM, Dooley J, Farr AG, Bevan MJ (2005) Notch ligands Delta 1 and Jagged1 transmit distinct signals to T-cell precursors. Blood 105:1440–1447

Leonard WJ (2001) Cytokines and immunodeficiency diseases. Nat Rev Immunol 1(3):200–208

Li L, Leid M, Rothenberg EV (2010) An early T cell lineage commitment checkpoint dependent on the transcription factor Bcl11b. Science 329:89–93

Li X, von Boehmer H (2011) Notch signaling in T-cell development and T-ALL. ISRN Hematol 2011:1–9

Love PE, Bhandoola A (2011) Signal integration and crosstalk during thymocyte migration and emigration. Nat Rev Immunol 11:469–477

Luc S, Luis TC, Boukarabila H, Macaulay IC, Buza-Vidas N, Bouriez-Jones T, Lutteropp M, Woll PS, Loughran SJ, Mead AJ, Hultquist A, Brown J, Mizukami T, Matsuoka S, Ferry H, Anderson K, Duarte S, Atkinson D, Soneji S, Domanski A, Farley A, Sanjuan-Pla A, Carella C, Patient R, de Bruijn M, Enver T, Nerlov C, Blackburn C, Godin I, Jacobsen SE (2012) The earliest thymic T cell progenitors sustain B cell and myeloid lineage potential. Nat Immunol 13:412–419

Maillard I, Fang T, Pear WS (2005) Regulation of lymphoid development, differentiation and function by the Notch pathway. Annu Rev Immunol 23:945–974

Maillard I, Tu L, Sambandam A, Yashiro-Ohtani Y, Millholland J, Keeshan K, Shestova O, Xu L, Bhandoola A, Pear WS (2006) The requirement for Notch signaling at the beta-selection checkpoint in vivo is absolute and independent of the pre-T cell receptor. J Exp Med 203:2239–2245

Maraskovsky E, O'Reilly LA, Teepe M, Corcoran LM, Peschon JJ, Strasser A (1997) Bcl-2 can rescue T lymphocyte development in interleukin-7 receptor-deficient mice but not in mutant rag-1-/- mice. Cell 89:1011–1019

Maraskovsky E, Peschon JJ, McKenna H, Teepe M, Strasser A (1998) Overexpression of Bcl-2 does not rescue impaired B lymphopoiesis in IL-7 receptor-deficient mice but can enhance survival of mature B cells. Int Immunol 10:1367–1375

Márquez C, Trigueros C, Fernández E, Toribio ML (1995) The development of T and non-T cell lineages from CD34+ human thymic precursors can be traced by the differential expression of CD44. J Exp Med 181:475–483

Márquez C, Trigueros C, Franco JM, Ramiro AR, Carrasco YR, López-Botet M, Toribio ML (1998) Identification of a common developmental pathway for thymic natural killer cells and dendritic cells. Blood 91:2760–2771

Michie AM, Carlyle JR, Schmitt TM, Ljutic B, Cho SK, Fong Q, Zúñiga-Pflücker JC (2000) Clonal characterization of a bipotent T cell and NK cell progenitor in the mouse fetal thymus. J Immunol 164:1730–1733

Mohtashami M, Shah DK, Nakase H, Kianizad K, Petrie HT, Zúñiga-Pflücker JC (2010) Direct comparison of Dll1- and Dll4-mediated Notch activation levels shows differential lymphomyeloid lineage commitment outcomes. J Immuno 185:867–876

Mungamuri SK, Yang X, Thor AD, Somasundaram K (2006) Survival signaling by Notch1: mammalian target of rapamycin (mTOR)-dependent inhibition of p53. Cancer Res 66: 4715–4724

Moore AJ, Sarmiento J, Mohtashami M, Braunstein M, Zúñiga-Pflücker JC, Anderson MK (2012) Transcriptional priming of intrathymic precursors for dendritic cell development. Development 139:373–384

Munitic I, Williams JA, Yang Y, Dong B, Lucas PJ, El Kassar N, Gress RE, Ashwell JD (2004) Dynamic regulation of IL-7 receptor expression is required for normal thymopoiesis. Blood 104:4165–4172

Pallard C, Stegmann AP, van Kleffens T, Smart F, Venkitaraman A, Spits H (1999) Distinct roles of the phosphatidylinositol 3-kinase and STAT5 pathways in IL-7-mediated development of human thymocyte precursors. Immunity 10:525–535

Palomero T, Lim WK, Odom DT, Sulis ML, Real PJ, Margolin A, Barnes KC, O'Neil J, Neuberg D, Weng AP, Aster JC, Sigaux F, Soulier J, Look AT, Young RA, Califano A, Ferrando AA (2006) NOTCH1 directly regulates c-MYC and activates a feed-forward-loop transcriptional network promoting leukemic cell growth. Proc Natl Acad Sci U S A 103:18261–18266

Palomero T, Sulis ML, Cortina M, Real PJ, Barnes K, Ciofani M, Caparros E, Buteau J, Brown K, Perkins SL, Bhagat G, Agarwal AM, Basso G, Castillo M, Nagase S, Cordon-Cardo C, Parsons R, Zúñiga-Pflücker JC, Dominguez M, Ferrando AA (2007) Mutational loss of PTEN induces resistance to NOTCH1 inhibition in T-cell leucemia. Nat Med 13:1203–1210

Pear WS, Aster JC, Scott ML, Hasserjian RP, Soffer B, Sklar J, Baltimore D (1996) Exclusive development of T cell neoplasms in mice transplanted with bone marrow expressing activated Notch alleles. J Exp Med 183:2283–2291

Peschon JJ, Morrissey PJ, Grabstein KH, Ramsdell FJ, Maraskovsky E, Gliniak BC, Park LS, Ziegler SF, Williams DE, Ware CB, Meyer JD, Davison BL (1994) Early lymphocyte expansion is severely impaired in interleukin 7 receptor–deficient mice. J Exp Med 180: 1955–1960

Petrie HT, Zúñiga-Pflücker JC (2007) Zoned out: functional mapping of stromal signaling microenvironments in the thymus. Annu Rev Immunol 25:649–679

Plum J, De Smedt M, Leclercq G, Verhasselt B, Vandekerckhove B (1996) Interleukin-7 is a critical growth factor in early human T-cell development. Blood 88:4239–4245

Porritt HE, Rumfelt LL, Tabrizifard S, Schmitt TM, Zúñiga-Pflücker JC, Petrie HT (2004) Heterogeneity among DN1 prothymocytes reveals multiple progenitors with different capacities to generate T cell and non-T cell lineages. Immunity 20:735–745

Puel A, Ziegler SF, Buckley RH, Leonard WJ (1998) Defective IL7R expression in T(-) B(+)NK(+) severe combined immunodeficiency. N. Nat Genet 20:394–397

Pui CH, Relling MV, Downing JR (2004) Acute lymphoblastic leukemia. N Engl J Med 350:1535–1548

Pui JC, Allman D, Xu L, DeRocco S, Karnell FG, Bakkour S, Lee JY, Kadesch T, Hardy RR, Aster JC, Pear WS (1999) Notch1 expression in early lymphopoiesis influences B versus T lineage determination. Immunity 11:299–308

Radtke F, Wilson A, Stark G, Bauer M, van Meerwijk J, MacDonald HR, Aguet M (1999) Deficient T cell fate specification in mice with an induced inactivation of Notch1. Immunity 10:547–558

Ramiro AR, Trigueros C, Márquez C, San Millán JL, Toribio ML (1996) Regulation of pre-T cell receptor (pT alpha-TCR beta) gene expression during human thymic development. J Exp Med 184:519–530

Rangarajan A, Talora C, Okuyama R, Nicolas M, Mammucari C, Oh H, Aster JC, Krishna S, Metzger D, Chambon P, Miele L, Aguet M, Radtke F, Dotto GP (2001) Notch signaling is a direct determinant of keratinocyte growth arrest and entry into differentiation. EMBO J 20:3427–3436

Reizis B, Leder P (2002) Direct induction of T lymphocyte-specific gene expression by the mammalian Notch signaling pathway. Genes Dev 16:295–300
Res P, Martínez-Cáceres E, Cristina Jaleco A, Staal F, Noteboom E, Weijer K, Spits H (1996) CD34+CD38dim cells in the human thymus can differentiate into T, natural killer, and dendritic cells but are distinct from pluripotent stem cells. Blood 87:5196–5206
Rich BE, Campos-Torres J, Tepper RI, Moreadith RW, Leder P (1993) Cutaneous lymphoproliferation and lymphomas in interleukin 7 transgenic mice. J Exp Med 177:305–316
Ronchini C, Capobianco AJ (2001) Induction of cyclin D1 transcription and CDK2 activity by Notch(ic): implication for cell cycle disruption in transformation by Notch(ic). Mol Cell Biol 21:5925–5934
Rothenberg EV (2007) Negotiation of the T-lineage fate decision by transcription factor interplay and microenvironmental signals. Immunity 26:690–702
Rothenberg EV, Moore JE, Yui MA (2008) Launching the T-cell-lineage developmental programme. Nat Rev Immunol 8:9–21
Sambandam A, Maillard I, Zediak VP, Xu L, Gerstein RM, Aster JC, Pear WS, Bhandoola A (2005) Notch signaling controls the generation and differentiation of early T lineage progenitors. Nat Immunol 6:663–670
Sánchez MJ, Muench MO, Roncarolo MG, Lanier LL, Phillips JH (1994) Identification of a common T/natural killer cell progenitor in human fetal thymus. J Exp Med 180:569–576
Schlenner SM, Madan V, Busch K, Tietz A, Läufle C, Costa C, Blum C, Fehling HJ, Rodewald HR (2010) Fate mapping reveals separate origins of T cells and myeloid lineages in the thymus. Immunity 32:426–436
Schluns KS, Kieper WC, Jameson SC, Lefrançois L (2000) Interleukin-7 mediates the homeostasis of na and memory CD8 T cells in vivo. Nat Immunol 1:426–432
Schmitt TM, Zúñiga-Pflücker JC (2002) Induction of T cell development from hematopoietic progenitor cells by delta-like-1 in vitro. Immunity 17:749–756
Schmitt TM, Ciofani M, Petrie HT, Zúñiga-Pflücker JC (2004) Maintenance of T cell specification and differentiation requires recurrent notch receptor-ligand interactions. J Exp Med 200:469–479
Shochat C, Tal N, Bandapalli OR, Palmi C, Ganmore I, te Kronnie G, Cario G, Cazzaniga G, Kulozik AE, Stanulla M, Schrappe M, Biondi A, Basso G, Bercovich D, Muckenthaler MU, Izraeli S (2011) Gain-of-function mutations in interleukin-7 receptor-α (IL7R) in childhood acute lymphoblastic leukemias. J Exp Med 208:901–908
Shortman K, Wu L (1996) Early T lymphocyte progenitors. Annu Rev Immunol 14:29–47
Shortman K, Egerton M, Spangrude GJ, Scollay R (1990) The generation and fate of thymocytes. Semin Immunol 2:3–12
Spits H (2002) Development of alphabeta T cells in the human thymus. Nat Rev Immunol 2(10):760–772
Stanley P, Guidos CJ (2009) Regulation of Notch signaling during T- and B-cell development by O-fucose glycans. Immunol Rev 230:201–215
Sudo T, Nishikawa S, Ohno N, Akiyama N, Tamakoshi M, Yoshida H, Nishikawa S (1993) Expression and function of the interleukin 7 receptor in murine lymphocytes. Proc Natl Acad Sci U S A 90:9125–9129
Taghon T, Van de Walle I, De Smet G, De Smedt M, Leclercq G, Vandekerckhove B, Plum J (2009) Notch signaling is required for proliferation but not for differentiation at a well-defined beta-selection checkpoint during human T-cell development. Blood 113:3254–3263
Takahama Y (2006) Journey through the thymus: stromal guides for T-cell development and selection. Nat Rev Immunol 6:127–135
Tan JB, Visan I, Yuan JS, Guidos CJ (2005) Requirement for Notch1 signals at sequential early stages of intrathymic T cell development. Nat Immunol 6:671–679
Thompson PK, Zúñiga-Pflücker JC (2011) On becoming a T cell; a converge of factors kick it up a Notch along the way. Semin Immunol 23:350–359
Trigueros C, Ramiro AR, Carrasco YR, de Yebenes VG, Albar JP, Toribio ML (1998) Identification of a late stage of small noncycling pTalpha- pre-T cells as immediate precursors of T cell receptor alpha/beta+ thymocytes. J Exp Med 188:1401–1412

Van de Walle I, De Smet G, De Smedt M, Vandekerckhove B, Leclercq G, Plum J, Taghon T (2009) An early decrease in Notch activation is required for human TCR-alphabeta lineage differentiation at the expense of TCR-gammadelta T cells. Blood 113:2988–2998

Van de Walle I, De Smet G, Gärtner M, De Smedt M, Waegemans E, Vandekerckhove B, Leclerq G, Plum J, Aster JC, Bernstein ID, Guidos CJ, Kyewski B, Taghon T (2011) Jagged2 acts as a Delta-like Notch ligand during early hematopoietic cell fate decisions. Blood 117:4449–4459

Van De Wiele CJ, Marino JH, Murray BW, Vo SS, Whetsell ME, Teague TK (2004) Thymocytes between the beta-selection and positive selection checkpoints are nonresponsive to IL-7 as assessed by STAT-5 phosphorylation. J Immunol 172:4235–4244

Varnum-Finney B, Brashem-Stein C, Bernstein ID (2003) Combined effects of Notch signaling and cytokines induce a multiple log increase in precursors with lymphoid and myeloid reconstituting ability. Blood 101:1784–1789

Vilimas T, Mascarenhas J, Palomero T, Mandal M, Buonamici S, Meng F, Thompson B, Spaulding C, Macaroun S, Alegre ML, Kee BL, Ferrando A, Miele L, Aifantis I (2007) Targeting the NF-kappaB signaling pathway in Notch1-induced T-cell leukemia. Nat Med 13:70–77

Visan I, Tan JB, Yuan JS, Harper JA, Koch U, Guidos CJ (2006) Regulation of T lymphopoiesis by Notch1 and Lunatic fringe-mediated competition for intrathymic niches. Nat Immunol:634-643

von Boehmer H, Fehling HJ (1997) Structure and function of the pre-T cell receptor. Annu Rev Immunol 15:433–452

von Boehmer H, Aifantis I, Azogui O, Feinberg J, Saint-Ruf C, Zober C, Garcia C, Buer J (1998) Crucial function of the pre-T-cell receptor (TCR) in TCR beta selection, TCR beta allelic exclusion and alpha beta versus gamma delta lineage commitment. Immunol Rev 165:111–119

von Freeden-Jeffry U, Vieira P, Lucian LA, McNeil T, Burdach SE, Murray R (1995) Lymphopenia in interleukin (IL)-7 gene-deleted mice identifies IL-7 as a nonredundant cytokine. J Exp Med 181:1519–1526

Wada H, Masuda K, Satoh R, Kakugawa K, Ikawa T, Katsura Y, Kawamoto H (2008) Adult T-cell progenitors retain myeloid potential. Nature 452:768–772

Wang H, Zou J, Zhao B, Johannsen E, Ashworth T, Wong H, Pear WS, Schug J, Blacklow SC, Arnett KL, Bernstein BE, Kieff E, Aster JC (2011) Genome-wide analysis reveals conserved and divergent features of Notch1/RBPJ binding in human and murine T-lymphoblastic leukemia cells. Proc Natl Acad Sci U S A 108:14908–14913

Washburn T, Schweighoffer E, Gridley T, Chang D, Fowlkes BJ, Cado D, Robey E (1997) Notch activity influences the alphabeta versus gammadelta T cell lineage decision. Cell 88:833–843

Weber BN, Wei-Shine Chi A, Chavez A, Yashiro-Ohtani Y, Yang Q, Shestova O, Bhandoola A (2011) A critical role for TCF-1 in T-lineage specification and differentiation. Nature 476:63–68

Weerkamp F, Baert MR, Brugman MH, Dik WA, de Haas EF, Visser TP, de Groot CJ, Wagemaker G, van Dongen JJ, Staal FJ (2006) Human thymus contains multipotent progenitors with T/B lymphoid, myeloid, and erythroid lineage potential. Blood 107:3131–3137

Weng AP, Ferrando AA, Lee W, Morris JP 4th, Silverman LB, Sanchez-Irizarry C, Blacklow SC, Look AT, Aster JC (2004) Activating mutations of NOTCH1 in human T cell acute lymphoblastic leukemia. Science 306:269–271

Weng AP, Millholland JM, Yashiro-Ohtani Y, Arcangeli ML, Lau A, Wai C, Del Bianco C, Rodriguez CG, Sai H, Tobias J, Li Y, Wolfe MS, Shachaf C, Felsher D, Blacklow SC, Pear WS, Aster JC (2006) c-Myc is an important direct target of Notch1 in T-cell acute lymphoblastic leukemia/lymphoma. Genes Dev 20:2096–2109

Wolfer A, Wilson A, Nemir M, MacDonald HR, Radtke F (2002) Inactivation of Notch1 impairs VDJbeta rearrangement and allows pre-TCR-independent survival of early alpha beta lineage thymocytes. Immunity 16:869–879

Wu L, Antica M, Johnson GR, Scollay R, Shortman K (1991) Developmental potential of the earliest precursor cells from the adult mouse thymus. J Exp Med 174:1617–1627

Xue HH, Bollenbacher J, Rovella V, Tripuraneni R, Du YB, Liu CY, Williams A, McCoy JP, Leonard WJ (2004) GA binding protein regulates interleukin 7 receptor alpha-chain gene expression in T cells. Nat Immunol 5:1036–1044

Yu Q, Erman B, Park JH, Feigenbaum L, Singer A (2004) IL-7 receptor signals inhibit expression of transcription factors TCF-1, LEF-1, and RORgammat: impact on thymocyte development. J Exp Med 200:797–803

Yu Q, Park JH, Doan LL, Erman B, Feigenbaum L, Singer A (2006) Cytokine signal transduction is suppressed in preselection double-positive thymocytes and restored by positive selection. J Exp Med 203:165–175

Zenatti PP, Ribeiro D, Li W, Zuurbier L, Silva MC, Paganin M, Tritapoe J, Hixon JA, Silveira AB, Cardoso BA, Sarmento LM, Correia N, Toribio ML, Kobarg J, Horstmann M, Pieters R, Brandalise SR, Ferrando AA, Meijerink JP, Durum SK, Yunes JA, Barata JT (2011) Oncogenic IL7R gain-of-function mutations in childhood T-cell acute lymphoblastic leukemia. Nat Genet 43:932–939

Notch Signaling During Human T cell Development

Tom Taghon, Els Waegemans and Inge Van de Walle

Abstract Notch signaling is critical during multiple stages of T cell development in both mouse and human. Evidence has emerged in recent years that this pathway might regulate T-lineage differentiation differently between both species. Here, we review our current understanding of how Notch signaling is activated and used during human T cell development. First, we set the stage by describing the developmental steps that make up human T cell development before describing the expression profiles of Notch receptors, ligands, and target genes during this process. To delineate stage-specific roles for Notch signaling during human T cell development, we subsequently try to interpret the functional Notch studies that have been performed in light of these expression profiles and compare this to its suggested role in the mouse.

Contents

T. Taghon (✉) · E. Waegemans · I. Van de Walle
Department of Clinical Chemistry, Microbiology and Immunology,
Ghent University Hospital, Ghent University, 9000 Ghent, Belgium
e-mail: tom.taghon@ugent.be

Current Topics in Microbiology and Immunology (2012) 360: 75–97

DOI: 10.1007/82_2012_230

Published Online: 14 July 2012

1 Human T cell Development

T cell development in postnatal life is a prolonged developmental process in which bone-derived multipotent hematopoietic progenitor cells seed the thymus (thymus seeding progenitors, TSPs) to become gradually reprogrammed into fully mature and functional T lymphocytes. The distinct developmental steps, as schematically illustrated in Fig. 1, are synchronized with the migration of the developing thymocytes towards specific niches in the thymus that provide the necessary stage-specific environmental factors that are needed for further differentiation (Petrie and Zuniga-Pflucker 2007). In recent years, much progress has been made with respect to the identification and characterization of these early T cell progenitors (ETPs) in human. Within the pool of intrathymic $CD34^{+}CD1a^{-}$ uncommitted T cell progenitors (population B, Fig. 1) (Blom and Spits 2006), Crooks and colleagues characterized three distinct subsets of progenitor cells that can be discriminated based on differential CD7 expression (Hao et al. 2008). The $CD34^{+}CD1a^{-}CD7^{-}$ subset seems to be the most immature subset of progenitors as it is mainly, but not exclusively, composed of $CD38^{-/low}$ progenitor cells, consistent with their potential to generate lymphoid, myeloid, and even erythroid cells (Hao et al. 2008; Weerkamp et al. 2006a). In contrast, $CD34^{+}CD1a^{-}CD7^{int}$ cells have lost myeloid and erythroid potential and thus resemble lymphoid primed progenitors that were earlier identified in umbilical cord blood by the same laboratory (Hao et al. 2001; Hoebeke et al. 2007). Both the $CD7^{-}$ and $CD7^{int}$ subset express CD10, and both populations have been identified in cord blood and adult bone marrow, raising the possibility that the human thymus can be colonized by both types of progenitor cells (population A, Fig. 1) (Six et al. 2007; Doulatov et al. 2010; Galy et al. 1995). While $CD34^{+}CD7^{int}CD10^{+}$ cells in the bone marrow have been proposed to be T-lineage committed prethymically (Klein et al. 2003), and while there is evidence from an in vitro xenograft model that only $CD7^{int}$ cells can colonize the thymus (Haddad et al. 2006), no in vivo models have been successfully used to demonstrate thymus homing from either subset (Six et al. 2007; Doulatov et al. 2010). In addition, the selective homing of $CD7^{int}$ cells does not fit with earlier experiments in which $CD34^{+}CD7^{-}$ progenitor cells were successfully used to study human T cell development using the same xenograft foetal thymus organ culture (FTOC) model (Taghon et al. 2002; Hoebeke et al. 2007), raising the possibility that few, but physiologically relevant and sufficient, $CD7^{-}$ cells can enter the thymus. Furthermore, it remains to be established whether both $CD7^{-}$ and $CD7^{int}$ cells enter as separate entities or if one subset leads to the development of the other. While it seems unlikely that $CD7^{int}$ cells give rise to more multipotent $CD7^{-}$ cells, the reverse cannot be excluded, especially since Notch activation results in induction of CD7 expression (Jaleco et al. 2001; De Smedt et al. 2002; Van de Walle et al. 2009; Magri et al. 2009; Van de Walle et al. 2011). Moreover, there is evidence that extrathymic $CD7^{int}$ progenitor cells still contain myeloid potential (Doulatov et al. 2010; Hoebeke et al. 2007), in contrast to the phenotypically similar population within the postnatal thymus (Hao et al. 2008).

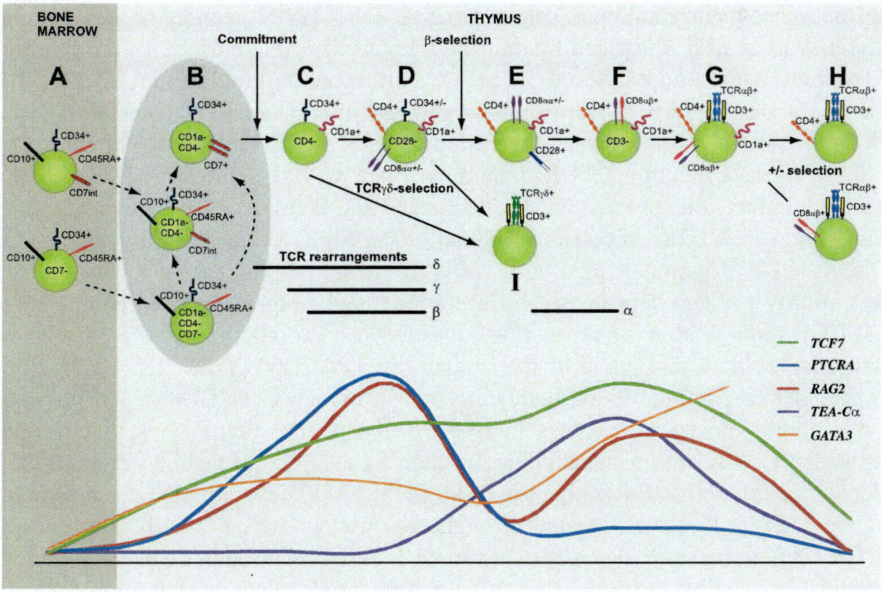

Fig. 1 Schematic overview of the different developmental stages that characterize human T cell development. Gene expression levels are indicative for the change in expression for each individual gene from one stage to the other, but do not provide insights into the differential expression levels between these genes. Data for *GATA3* in population H are not provided due to the high difference in expression between CD4 and CD8 $\alpha\beta$ T cell populations

However, the apparent discrepancies might be due to the technical approaches used. Since sustained Notch signaling is required to suppress B cell development (Taghon et al. 2005; Krueger et al. 2006), it is possible that CD34$^+$CD7^{int} progenitors have just initiated the T-lineage specification program as they still display B-lineage differentiation potential in vitro. Further studies using clonal approaches will be required to fully resolve this issue.

Within the CD34$^+$CD1$^-$ uncommitted pool of progenitors, the third—and by far largest—subset of thymocytes expresses high levels of CD7 and mainly comprises T/NK progenitors, resembling DN2a cells in the mouse (Yui et al. 2010). During postnatal life, these cells are most likely derived from Notch-primed CD34$^+$CD1a$^-$CD7$^-$ and/or CD34$^+$CD1a$^-$CD7^{int} cells as they do not seem present within the bone marrow or peripheral blood (our own unpublished observations). In addition, these cells display high expression levels of T-lineage specific genes and have faster T-lineage kinetics compared to the other uncommitted subsets, indicating that these cells have been specified toward the T-cell lineage (Hao et al. 2008; Van de Walle et al. 2009). Further differentiation induces T-cell commitment which is complete when the immature T cell marker CD1a is expressed (population C, Fig. 1) (Blom and Spits 2006).

During these specification and commitment processes, TCR rearrangements at the *TCRD*, *TCRG* and *TCRB* loci are initiated (Dik et al. 2005), and are fully active in the subsequent immature single positive CD4 (ISP4) subset of developing thymocytes as also illustrated by their high levels of *RAG* and *PTCRA* (coding for pTα) expression (population D, Fig. 1), corresponding to murine DN3a cells (Taghon et al. 2006). In-frame rearrangements that yield a TCR-δ and TCR-γ chain will mainly result in the generation of TCR$\gamma\delta^+$CD3$^+$ T cells (population I, Fig. 1), while a functional TCR-β chain will pair with the surrogate TCRα chain, pTα, to induce the process of β-selection. This event is characterized by the acquisition of CD28 expression (population E, Fig. 1, corresponding to murine DN3b cells) (Taghon et al. 2006, 2009; Blom and Spits 2006), extensive but temporarily Notch-dependent proliferation and rapid differentiation into CD4$^+$CD8$\alpha\beta^+$ double positive (DP) thymocytes (population F, Fig. 1). It is important to note that some of the immature CD4$^+$CD3$^-$CD28$^-$ ISP4 cells also express CD8$\alpha\alpha$ homodimers, indicating that in human CD4$^+$CD8$\alpha\alpha^+$ DP thymocytes do not necessarily represent true $\alpha\beta$-lineage DP thymocytes that passed through the β-selection checkpoint (Taghon et al. 2009; Joachims et al. 2006; Carrasco et al. 1999). Subsequently, CD4$^+$CD8$\alpha\beta^+$ DP cells initiate rearrangement of the TCR-α chain, as revealed through the onset of sterile T-early α transcript (TEA-Cα) expression (Fig. 1), to generate a fully functional TCR-$\alpha\beta$ complex (population G, Fig. 1). Finally, positive and negative selection determines which cells further mature into CD4$^+$ or CD8$^+$ SP TCR$\alpha\beta$ T cells (population H, Fig. 1) (Plum et al. 2008).

2 Notch Signaling Profile

As in the mouse, the Notch signaling pathway is involved in various stages of T cell development in human. Given the fact that the Notch signaling pathway is composed of several receptors and ligands that can activate a broad range of different target genes, we will first describe our current understanding of the expression patterns of these components during human T cell development. This will serve as point of reference for interpreting the functional studies that have been performed thus far.

Migration through the thymus is critical for developing thymocytes to receive the appropriate stage-specific developmental cues for further differentiation. As a result, characterization of the expression of Notch ligands by stromal cells at specific sites of the thymus will provide a hint of the possible involvement of these Notch signal initiating components during human T cell development. In collaboration with the Kyewski lab (Gotter et al. 2004), we recently determined the expression pattern of Notch ligands in human cortical and medullary thymic epithelial cells (cTEC and mTEC, respectively) using quantitative PCR (Q-PCR) and flow cytometry, thereby providing insights into the Notch ligands that possibly support early stages of human thymocyte development (as shown by expression in cTECs) and those that may be involved in the final maturation stages of T cell development (as shown by expression in mTECs) (Van de Walle et al. 2011). Both

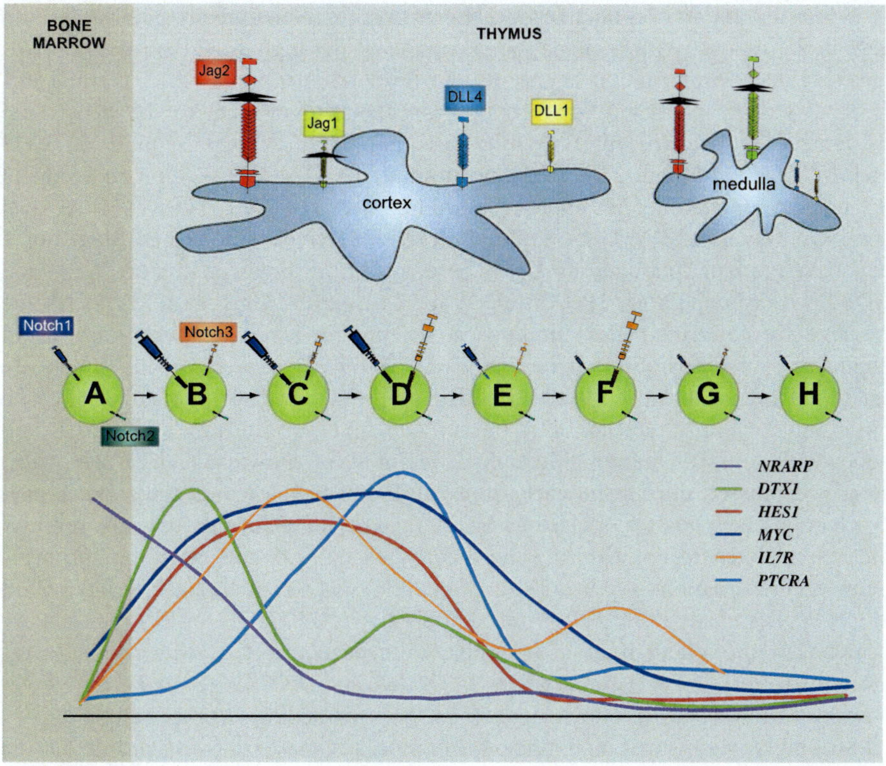

Fig. 2 Overview of the expression of the notch signaling components during human T cell development. The size of the notch ligands and receptors is a measure for their expression level. The expression levels for the notch downstream target genes are indicative for the change in expression for each individual target gene from one stage to the other, but do not provide insights into the differential expression levels between these genes. Data for *IL7R* in population H are not provided due to the high difference in expression between CD4 and CD8 $\alpha\beta$ T cell populations

approaches revealed a predominant expression of *JAG2*, in 70–90 % of both TEC subsets as defined using flow cytometry. Between 10 and 20 % of cTECs expressed *DLL4*. In contrast, only very low amounts of Delta-Like-1 protein were detected in no more than 10 % of cTECs and *JAG1* was mainly expressed by mTECs, as revealed through Q PCR only due to lack of suitable antibodies. A schematic overview of these expression patterns is presented in Fig. 2. From this, one could predict that Jagged2 might play a major role during human T cell development. Unfortunately, the generation of a Jagged2 deficient human thymic microenvironment is currently impossible, although approaches using hES cells might become feasible in the future (Green et al. 2011). As discussed further, the specific expression of *DLL4* in cTECs is consistent with its requirement for the induction of T-lineage differentiation in TSPs, as is the medullary expression of the weak Notch ligand *JAG1* as Notch signaling does not seem to play an obvious role in the final maturation stages of T cell development.

While the expression data for the Notch ligands seems mostly consistent with data in the mouse, with perhaps the exception of the high *Jag2* expression levels that still need confirmation at the protein level (Heinzel et al. 2007), some differences can be observed between both species with respect to the expression patterns of the Notch receptors. While TSPs (population A) express both *NOTCH1* and *NOTCH2*, but not *NOTCH3*, uncommitted CD34^{+}CD1^{-} human postnatal thymocytes (population B) express, in addition to *NOTCH1* and *NOTCH2*, also significant levels of *NOTCH3*, and this expression persists into the DP stages of T cell development (populations C–G) before shutting down in mature CD4 and CD8 SP cells (population H) (Van de Walle et al. 2009; Ghisi et al. 2011). While we recently confirmed these findings at the protein level (Van de Walle et al. manuscript in preparation), the kinetics of *NOTCH3* expression in human are earlier compared to those observed during mouse T cell development, raising the possibility that this receptor may be actively involved during the initial Notch dependent stages of human T-lineage differentiation. However, *NOTCH1* is also clearly expressed during the early steps of T cell development until the human β-selection checkpoint (populations B–D), raising the possibility that this remains the major Notch receptor during these early stages of T cell development, analogous to the situation in mouse (Suliman et al. 2011; Shi et al. 2011; Radtke et al. 1999; Wolfer et al. 2002; Wilson et al. 2001). Also *NOTCH2* mRNA can clearly be detected throughout T cell development (Van de Walle et al. 2009), but at levels around 10-fold lower compared to *NOTCH1* and *NOTCH3*, predicting lower chances to interact with Notch ligands on the surrounding stromal cells.

Strikingly, we observed significant differences in the expression profiles of the major Notch target genes *NRARP*, *DTX1*, *HES1*, *MYC* and *PTCRA* during human T cell development (illustrated graphically in Fig. 2) (Taghon et al. 2009; Van de Walle et al. 2009). *NRARP* is the only Notch target gene that is already expressed in extrathymic progenitors (population A) and its expression is maintained in uncommitted postnatal thymocytes (population B) before declining upon T cell commitment (population C). In contrast, *DTX1* is absent in TSPs (population A) but is specifically upregulated in early uncommitted T cell progenitors (population B) before immediately declining again upon T-lineage commitment (population C). At present, it is unclear which CD34^{+}CD1^{-} subpopulations are responsible for *DTX1* expression, but such information will be of interest for characterizing the cells that receive the first Notch triggers (Sambandam et al. 2005). A smaller, second wave of *DTX1* expression can be observed just before β-selection (population D), in DN3a-like cells that express high levels of another Notch target gene *PTCRA*, which appropriately is only expressed in these TCR rearranging cells, awaiting the functional production of a TCR-β chain to induce β-selection and further differentiation. While *HES1* and *MYC* are also induced upon initiation of T cell development in CD34^{+}CD1^{-} uncommitted thymocytes, their expression, in contrast to *DTX1*, is maintained in CD34^{+}CD1^{+} committed and subsequent T cell progenitors until the cells receive a preTCR signal and pass through the β-selection checkpoint. Following this β-selection checkpoint, no significant expression of any of these Notch downstream targets can be detected, consistent with the requirement for the initiation of *TCRA* gene rearrangements as documented through the induction of expression of sterile TEA-Cα transcripts, a process that is inhibited through Notch signaling (Van de Walle et al. 2009; Taghon et al. 2009). The only exception to this is *NOTCH3*, a gene that can be upregulated downstream of Notch1 signaling (Palomero et al. 2006; Weng et al. 2006; Van de Walle et al. 2011). In contrast to *NOTCH1*, *NOTCH3* expression is again upregulated following β-selection before being shut off in mature SP CD4 and CD8 cells.

Although we could show that *HES1*, *DTX1*, *NRARP*, *MYC*, *PTCRA,* and *NOTCH3* are also Notch dependent in human thymocytes, albeit to different degrees (Van de Walle et al. 2009; Taghon et al. 2009), their highly diverse expression profiles suggest that additional regulatory mechanisms must be involved in controlling their expression. *IL7R* was also suggested to be a direct Notch target during human T cell development (Gonzalez-Garcia et al. 2009; Magri et al. 2009), but nobody thus far has been able to show any short-term Notch-dependent effects on *IL7R* expression in primary human thymocytes [(Gonzalez-Garcia et al. 2009; Magri et al. 2009) and our own unpublished observations]. In addition, this gene also displays a unique expression pattern during human T cell development that is distinct from any of the other Notch target genes (Fig. 2). As suggested initially, differential Notch-dependency of these downstream target genes may partially explain their differential expression patterns; however, it cannot account for all of the observed differences as, for instance the profiles of *DTX1* and *NRARP*, on the one hand, and that of *PTCRA*, on the other hand, are too distinct. This is also exemplified by *NOTCH3*, which can be a Notch1 downstream target gene, especially during the early stages of T cell development (Van de Walle et al. 2011; Weerkamp et al. 2006b; Neves et al. 2006), but the absence of expression of any other Notch downstream target gene, as well as of the Notch1 receptor itself at that same developmental stage, indicate that other molecular mechanisms are driving this expression. The need for additional regulatory inputs, such as feed-forward mechanisms that take over following a Notch-dependent initial induction, is also obvious from the fact that Notch signaling by itself, as provided in an OP9 coculture system through excess expression of *DLL1*, is not sufficient to maintain the same expression levels of most of these genes compared to their ex vivo isolated counterparts (Van de Walle et al. 2009). The exceptions to this are *DTX1* and *NRARP*, not accidently the most Notch-dependent target genes that we could find (Van de Walle et al. 2009; Taghon et al. 2009). Thus, it remains to be investigated if perhaps auto-regulatory mechanisms, as documented for *HES1* (Hirata et al. 2002), or other transcriptional regulators, as also illustrated previously in the mouse (Ikawa et al. 2006), regulate these processes. Given the recent finding that mir-150 is involved in turning off *NOTCH3* expression upon human T cell maturation (Ghisi et al. 2011), it will also be critical to investigate the involvement of non-coding RNAs or other epigenetic phenomena in these settings.

Strikingly, with the exception of the apparent global silencing of Notch target genes following β-selection, the differential expression of the human Notch target genes is very distinct from what is observed during mouse T cell differentiation as all of these genes are gradually upregulated during the initial stages of murine T cell development before peaking at the DN3a stage (Taghon et al. 2006; David-Fung et al. 2006; Tydell et al. 2007; Weng et al. 2006). These findings suggest that the Notch signaling pathway, and its individual components, are used differently during T cell development in both species. As discussed below, we believe that at least some of these apparent discrepancies help to explain some of the functional differences that have been observed with respect to the role of Notch signaling during mouse and human T cell development since the same target genes have to integrate in different stage-specific regulatory networks.

3 Stage-Specific Notch Signaling Requirements

The use of conditional knockout mice has significantly advanced our understanding of the role of these individual Notch signaling components during mouse T cell development. However, such an approach has not been easily accessible thus far in a human setting, with the exception of the recently generated Notch receptor specific antibodies that have not been widely used yet (Wu et al. 2010; Li et al. 2008). In anticipation also of gene-targeting studies in human embryonic stem cells (Hockemeyer et al. 2009, 2011) that can subsequently be used for studying the function of any desired gene during T cell development (Vandekerckhove et al. 2011; Timmermans et al. 2009; Galic et al. 2006), human Notch studies have been limited to the use of less specific tools, such as pharmacological γ-secretase inhibitors (GSIs), or 'all or nothing' approaches, for instance, using stromal cocultures in the presence or absence of a Notch ligand and overexpression studies using intracellular Notch (*ICN*) or the dominant-negative mutant of Mastermind-like-1 (*DNMAML1*). We will now discuss the functional implementation of these studies in light of the expression patterns of the individual Notch signaling components at specific stages of human T cell development.

3.1 Induction Of T-Lineage Specification

In the mouse, the Delta-Like-4/Notch1 interaction is considered to be the main driving force for the Notch-dependent initiation of T cell development since both *Notch1* and *Dll4* conditional deletion studies have provided unambiguous evidence for this (Koch et al. 2008; Radtke et al. 1999; Wilson et al. 2001; Hozumi et al. 2008; Feyerabend et al. 2009), in contrast to, for instance, DLL1 (Hozumi et al. 2004) or Notch3 (Suliman et al. 2011; Shi et al. 2011) knockout approaches. Absence of either *Notch1* or *Dll4* inhibits T cell development and results in differentiation of other hematopoietic lineages instead, such as B, NK and myeloid cells. In human, it is clear from *ICN* (De Smedt et al. 2002) and *DNMAML1* (Taghon et al. 2009) overexpression experiments, GSI inhibition data (De Smedt et al. 2005; Van de Walle et al. 2009) and stromal cocultures data (De Smedt et al. 2004; La Motte-Mohs et al. 2005; Lefort et al. 2006; Benne et al. 2009; Awong et al. 2009; Van de Walle et al. 2011) that Notch is equally important in humans to induce T cell development at the expense of other hematopoietic cell types. In this setting, Notch seems to act as a rheostat in which graded levels of signaling affect the hematopoietic lineage outcome (De Smedt et al. 2005; Lefort et al. 2006; Jaleco et al. 2001). Both functional studies using GSI (De Smedt et al. 2005; Van de Walle et al. 2009) and the Notch target gene expression profile [Fig. 2 (Van de Walle et al. 2009; Taghon et al. 2009)] suggest that a strong Notch signal is required for driving the T-lineage specification process (Van de Walle et al. 2009). *DTX1*, *HES1* and *MYC* are specifically upregulated in uncommitted $CD34^{+}CD1^{-}$ thymocytes compared to extrathymic progenitors, while *NRARP* expression is maintained (Fig. 2). Given that only *NOTCH1* and *NOTCH2* seem to be expressed by TSPs, and that Delta-Like-4-mediated Notch1 activation induces a stronger Notch signal compared to when induced by Jagged2, it seems very likely that the Delta-Like-4/Notch1 interaction is also responsible for initiating human T-lineage specification, despite the abundant *JAG2* expression by human TECs and the potential of the protein to induce, and support T cell development in human hematopoietic progenitor cells (Van de Walle et al. 2011).

3.2 Induction Of T-Cell Commitment

Interestingly, following this strong Notch signal that induces the T-cell specifying transcriptional program (Taghon et al. 2005; Weerkamp et al. 2006a; Van de Walle et al. 2011), a reduction in the expression of the target genes *DTX1* and *NRARP*, but not *HES1* and *MYC*, is observed in human thymocytes as they become T-lineage committed progenitors. Since those two genes are very sensitive to small changes in Notch signaling intensities (Van de Walle et al. 2009), this suggests that Notch signal strength is reduced during this transition. In addition, given that both Nrarp and Deltex1 are considered to be negative regulators of the Notch signaling pathway, silencing of the genes that code for these proteins might be sufficient, and perhaps critical, to allow further Notch-dependent human T cell differentiation, during which the Notch signal is still sufficiently strong to allow expression of other Notch targets. Consistent with that idea, various experiments from different labs have shown that a reduction in Notch signal strength in uncommitted $CD34^{+}CD1^{-}$ human T-lineage progenitor cells allows and perhaps even enhances further differentiation into DP thymocytes (Van de Walle et al. 2009; Magri et al. 2009; Dontje et al. 2006). While our lab used graded dosages of GSI in the OP9-DLL1 coculture system, similar results were obtained when human $CD34^{+}CD1^{-}$ uncommitted thymocytes were plated on OP9 cells expressing *JAG1* (Dontje et al. 2006), the weakest Notch1 ligand (Van de Walle et al. 2011) but with the potential to induce *HES1* expression (Neves et al. 2006; Van de Walle et al. 2011). In such conditions of weakened Notch activation, human T-lineage progenitor cells can differentiate fully into CD3$^+$TCR$\alpha\beta^+$ thymocytes [(Dontje et al. 2006) and our own unpublished results]. Given that *JAG2*, but not *JAG1*, is abundantly expressed by human cTECs that are mediating these stages of T cell development, and given that Jagged2 is a weaker activator of Notch1 compared to Delta-Like-4, a change in ligand-mediated Notch1 activation from Delta-Like-4 to Jagged2 might be responsible for reducing Notch activation upon T cell commitment in vivo. Alternatively, a local reduction in Delta-Like-4 density may also be regulating this process. Importantly, one also has to consider that the additional presence of the Notch3 receptor in these cells may alter the balance of Notch signaling activity. While Delta-Like-4 does not seem to be a good ligand for this receptor in the mouse (Suliman et al. 2011), it remains to be investigated whether Delta-Like-4 and/or Jagged2 can bind and activate Notch3 in humans. This is, particularly, important since Notch3 has been suggested to be a negative regulator of Notch1 activation (Beatus et al. 1999), suggesting that Notch3 activation might be another mechanism to down-modulate Notch1 activity. Thus, multiple changes in Notch receptor/ligand interactions can occur that might regulate the reduction in Notch signaling activity that seems involved in supporting the further development of T-lineage specified human T cell progenitors.

Strikingly, even in the absence of Notch signaling, uncommitted human thymocytes can generate CD8β expressing DP cells—even CD3$^+$TCR-$\alpha\beta^+$ cells when provided with a rearranged TCR-β chain, although very inefficient due to lack of Notch-dependent proliferation and/or survival (Taghon et al. 2009). While we do not wish to imply that Notch signaling is no longer involved in further T cell differentiation (as also discussed further in this review), it does illustrate that other molecular mechanisms might be more important for driving further developmental progression. This may involve the activity of critical T-lineage transcription factors such as TCF-1, GATA-3, and BCL11b (Verbeek et al. 1995; Weber et al. 2011; Ting et al. 1996; Taghon et al. 2001, 2007; Hosoya et al. 2009; Li et al. 2010a, b; Ikawa et al. 2010). Such mechanisms might also be recruited to silence the Notch target genes *DTX1* and *NRARP*, while the Notch signaling activity itself, as measured by the amount of ICN that is released, remains at its initial signaling strength. This could be another mechanism through which Notch target genes are regulated differentially. While this has to be investigated further, the function of these Notch target genes during T cell development also remains unclear. Understanding this should also help to explain their specific expression profile at this early stage of human T cell development. In each case, Notch does not seem to be essential to induce human T-cell commitment in previously T-lineage specified progenitors in the absence of exogenous, non-T-lineage, cytokines that can drive alternative lineage potential (De Smedt et al. 2005, 2007; Taghon et al. 2009; Magri et al. 2009). Given that the only function of the thymus is to give rise to T cells, it rather seems that the high initial levels of Notch signaling in human are important to expand the initial pool of uncommitted T-lineage specified progenitors that are generated from the limited number of TSPs.

Strikingly, the peak of *DTX1* and *NRARP* expression in uncommitted human thymocytes is in sharp contrast to the gradual increase in expression that is observed for all Notch target genes up to the DN3a stage during mouse T cell development. Consistent with that, also functional studies with mouse T-lineage progenitors reveal that strong Notch signaling remains essential for inducing murine T cell commitment and further differentiation into DP $\alpha\beta$-lineage cells. Reduction (Lehar et al. 2005) or complete inhibition (Schmitt et al. 2004; Feyerabend et al. 2009) of Notch signal strength results in alternative lineage differentiation or cell death when initiated with uncommitted progenitor cells. It is intriguing that this 'lock-down' of T-lineage commitment in mouse and human is so differentially dependent on Notch signaling. While it is difficult to compare Notch signaling intensities across species, one explanation could be that the required signaling thresholds to mediate these events are different between mouse and human.

3.3 TCR Rearrangements, β-Selection and Beyond

Importantly, Notch signaling remains essential following human T cell commitment to support TCR rearrangements and proliferation of the cells. At this point in development, thymocytes start to express the necessary genes to allow gene recombination at the *TCRD*, *–G* and*–B* loci, such as *IL7R* and the *RAG* genes, as well as *PTCRA* to allow immediate induction of preTCR signaling and subsequent β-selection upon the generation of an in-frame TCR-β chain. Consistent with a clear requirement for Notch signaling to support human TCR-β chain rearrangements (De Smedt et al. 2005), also *PTCRA* expression seems Notch dependent, although we could only observe a small reduction in *PTCRA* expression upon Notch inhibition in human thymocytes using GSI (Van de Walle et al. 2009) or *DNMAML1* overexpression (Fig. 3). While a conserved CSL binding site has been detected in both human and mouse (Reizis and Leder 2002), *PTCRA* expression in human seems at least equally dependent on other regulatory inputs since its expression drops 7-fold in conditions of excess Notch stimulation compared to ex vivo isolated cells (Van de Walle et al. 2009). Somewhat inconsistent with this critical role for Notch in the generation of a human preTCR complex is the notion that expression of *RAG1* and *-2*, genes equally required to generate a functional rearranged TCR-β chain, is enhanced upon removal of Notch signaling [Fig. 3 and (Van de Walle et al. 2009)]. In our current experimental setting, this may, however, also reflect the requirement for removal of Notch signaling following β-selection to allow TCRα gene rearrangements. Whether or not Notch is involved in TCR rearrangements of the *TCRD* and *–G* gene segments is presently unclear. As such, the precise role for Notch signaling during all of these recombination processes remains to be established. It is intriguing to observe that *NOTCH3* displays a very similar expression profile as both *RAG* genes, suggesting that this receptor might be involved in their regulation. As mentioned above, since Notch1 and Notch3 have been proposed to play opposing roles with respect to the activation of Notch target genes (Beatus et al. 1999), the interplay between both receptors might

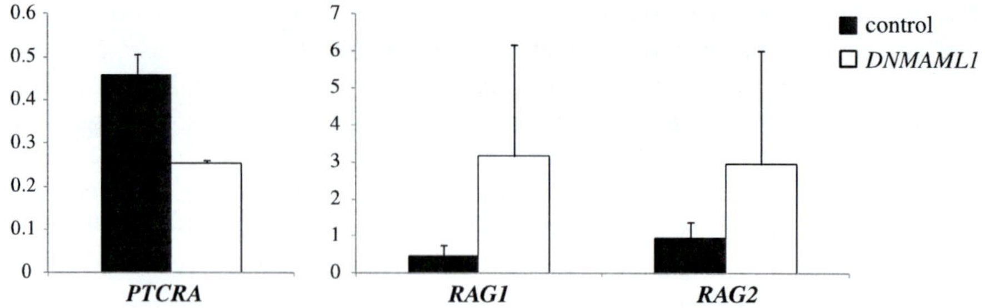

Fig. 3 Inconsistent Notch-dependent regulation of genes critical for preTCR formation during human T cell development

be involved in the apparent differential Notch dependency of for instance *PTCRA*, on the one hand, and *RAG1* and *RAG2*, on the other hand. While both assumingly compete for interaction with CSL, they might attract other co-activators that differentially affect the transcriptional activity of each Notch receptor specific activation complex. Notch1 and Notch3 differ with respect to their transactivation domain but little is known about its functional importance. Chromatin-immunoprecipitation studies should be able to provide some insights into this in the near future.

During the rearrangement processes, thymocytes do not cycle due to a high risk of DNA damage that could be induced by the recombinase activity of the RAG proteins that are present in the cells at that particular stage. As such, repression of *RAG* expression through Notch signaling makes sense as Notch activity generally induces proliferation in immature thymocytes. Whether or not Notch signaling is required for survival of the TCR rearranging cells during these stages has not been thoroughly investigated in humans. While it is clear that Notch signaling is critical for the long-term survival of human thymocytes, it is not well-established whether this is mediated directly through Notch signaling events that regulate glucose metabolism as shown for mouse DN3 thymocytes (Ciofani and Zuniga-Pflucker 2005), or rather indirectly due to the lack of other critical survival signals such as, for instance, provided through TCR signaling (De Smedt et al. 2005). Expression of *MYC*, coding for an important transcriptional regulator of genes involved in cellular growth and survival, does seem to be Notch-dependent during these stages of human T cell development, suggesting that Notch at least partially regulates these cellular processes during early human thymocyte development. However, while a slight increase in the frequency of apoptotic cells has been observed upon Notch removal in the OP9 stromal coculture system (Magri et al. 2009), such an increase in cell death is less obvious when a more physiological system with a thymic microenvironment is used, suggesting that other survival signals besides Notch might prove to be more important (De Smedt et al. 2005). Given that the mouse experiments that revealed a requirement for Notch signaling in regulating the glucose metabolism were performed in the OP9 coculture system (Ciofani and

Zuniga-Pflucker 2005), it would be of interest to confirm these findings in a thymic environment. Intriguingly, Notch1 deletion in vivo at the DN2 to DN3 transition of mouse T cell development does permit survival of preTCR deficient $\alpha\beta$-lineage DN4-like cells that resemble post β-selection thymocytes, suggesting that other signals beside Notch mediate in vivo survival of these immature cells (Wolfer et al. 2002). On the other hand, complete inhibition of canonical Notch signaling using continuous *DNMAML1* expression in DN3 thymocytes in vivo does reveal a more severe effect on early thymocyte survival (Maillard et al. 2006). It seems, unlikely, that the differential outcome between both approaches involves any Notch3 activation that could provide compensatory Notch signals (Shi et al. 2011). Nevertheless, the critical role that Notch signaling seems to play at this stage of mouse T cell development, together also with its requirement to support TCR-β rearrangements (Wolfer et al. 2002) and to positively regulate *PTCRA* expression, fits with the peak in Notch target gene expression that is observed in these DN3a thymocytes. Thus, if we consider the peak of Notch signaling intensity, determined by the expression level of downstream target genes, as a measurement for deciphering the stage at which Notch activity might play its most significant role during T cell development, this might help to explain why this pathway seems to play a less critical role in the survival of early human thymocytes, compared to in mouse.

Also at the β-selection checkpoint, Notch does not seem as stringently required for the differentiation of human thymocytes compared to in mouse. In the absence of Notch signaling prior to successful TCR-β rearrangement, thymocytes that receive an exogenously provided TCR-β chain will further differentiate into CD8β^+ DP thymocytes that will subsequently rearrange the TCR-α chain to become TCR-$\alpha\beta^+$CD3$^+$ cells (Taghon et al. 2009). While Notch does seem critical for appropriate proliferation of the cells in the OP9 coculture system—we cannot exclude that this apparent Notch-dependent proliferation is a result of lack of other proliferative signals that are present in vivo—such further differentiation in the absence of Notch has not been observed in the mouse (Ciofani et al. 2004; Taghon et al. 2006; Garbe et al. 2006). Consistent with above, this may also be the result of a more robust survival of human thymocytes in the absence of Notch activity. In each case, Notch signaling does not seem to be as stringently required during human as compared to during mouse β-selection.

Following β-selection, there is no evidence at present that Notch activity is critically required for the further differentiation of human thymocytes into functionally mature T cells, but this has not been thoroughly investigated yet. While *NOTCH3*, but not *NOTCH1*, is specifically upregulated in human DP thymocytes, none of the well-characterized Notch target genes seem to be induced, suggesting that this receptor is either not activated, or that it does not have the potential to activate these genes (Beatus et al. 1999). Further work, in which *NOTCH3* can be targeted specifically, will be required to determine if this receptor plays a functional role at this stage of human T cell development.

4 TCR-$\alpha\beta$ Versus TCR-$\gamma\delta$ Lineage Choice

Besides the main population of TCR-$\alpha\beta$ T cells that differentiate within the thymus, a subset of T cells expressing TCR-$\gamma\delta$ also develops during intrathymic T cell differentiation, although these cells display more innate-like lymphocyte properties. Although $\gamma\delta$ T cells also circulate within the blood like their $\alpha\beta$-lineage counterparts, they probably have a more prominent role in mucosa-associated tissues within the body. In contrast to TCR-$\alpha\beta$ T cells that belong to the adaptive immune system and have the potential as a family to recognize virtually all possible antigens through their high diversity in TCRs, $\gamma\delta$ T cells recognize more structural and less-diverse antigens. As mentioned above, $\gamma\delta$ T cells branch of from the 'mainstream' $\alpha\beta$-lineage pathway during the $CD34^{+}CD1^{+}$ and $CD4^{+}CD8\alpha\alpha^{-/+}CD28^{-}$ stages of human T cell development when *TCRD*, *-G* and *-B* gene segments rearrange (Van de Walle et al. 2009; Joachims et al. 2006). Whether they differentiate through a $CD28^{+}$ or $CD71^{+}$ stage prior to TCR expression is presently unclear. Since newly generated $\gamma\delta$ T cells are not as proliferative as preTCR selected $\alpha\beta$-lineage thymocytes, such a transient stage might be difficult to identify, also because the full TCR complex is presumably immediately expressed.

Human $\gamma\delta$ T cells preferentially develop in conditions of high Notch activation and this was initially demonstrated using *ICN* overexpression studies (De Smedt et al. 2002; Garcia-Peydro et al. 2003). Since these results were inconsistent with the proposed less critical role for Notch in mouse $\gamma\delta$ T cell development (Robey et al. 1996; Doerfler et al. 2001), it raised the issue whether these superficial Notch signaling levels provided any insights into the normal physiological role of Notch in this developmental process in human. Importantly, OP9 coculture experiments with human progenitors later confirmed these findings and showed, not only that high Notch signaling favors $\gamma\delta$ over $\alpha\beta$-lineage differentiation (Van de Walle et al. 2009), but also that virtually no $\gamma\delta$ T cells can develop from human T-lineage specified or T-cell committed progenitors in the absence of Notch signaling (Taghon et al. 2009). Since such an approach was more compatible with experiments that had been performed with mouse T-lineage precursors, this provided more robust insights into the differential Notch signaling requirements between mouse and human $\gamma\delta$ T cells. Consistent with the hypothesis that Notch signaling is most critical for human $\gamma\delta$ T cell development, we did observe a slightly increased expression of the Notch target genes *DTX1*, *HES1*, *NRARP* and *MYC* in $\gamma\delta$- compared to $\alpha\beta$-lineage cells, although this level was still reduced compared to in early $CD34^{+}CD1^{-}$ human thymocytes (Van de Walle et al. 2009). In addition to these genes, *RUNX3* showed an even more interesting expression pattern. While *RUNX3* expression is high in $CD34^{+}CD1^{-}$ uncommitted thymocytes and downregulated upon the induction of T-cell commitment in $CD34^{+}CD1^{+}$ thymocytes, *RUNX3* levels reach an equally high level in $\gamma\delta$-lineage cells compared to uncommitted $CD34^{+}CD1^{-}$ thymocytes. While there is currently no direct evidence that this gene is a direct Notch target gene, GSI titration experiments did suggest some degree of Notch dependent regulation in $CD34^+$ human postnatal thymocytes (Van de Walle et al. 2009), raising the possibility that RUNX3 mediates the Notch driven human $\gamma\delta$-lineage differentiation.

The choice of early T-lineage specified progenitors to develop along the $\gamma\delta$- or $\alpha\beta$-differentiation pathways has been extensively studied in the mouse over the past years and has been recently reviewed in depth (Ciofani and Zuniga-Pflucker, 2010; Kreslavsky et al. 2010; Lee et al. 2010; Taghon and Rothenberg 2008). Overall, the consensus is that the outcome of TCR rearrangements is not the sole determinant of the final lineage outcome. Following rearrangements, signaling through the generated $\gamma\delta$- or preTCR can still alter the TCR-predicted lineage outcome as a strong TCR-signal will drive further differentiation along the DN $\gamma\delta$-pathway, while a weaker TCR signal will promote DP $\alpha\beta$-lineage differentiation. Downstream of the TCR, Id3 is a critical mediator that translates the strength of the TCR signal into a developmental lineage choice by modulating E protein activity (Lee et al. 2010). However, other signaling pathways have been shown to integrate with TCR signaling to impact the resulting lineage outcome, also through altering E protein activity. In this context, strong Notch signals have been shown to be essential to promote $\alpha\beta$-lineage differentiation in the mouse, as characterized by the generation of DP thymocytes, following preTCR signaling. In contrast, $\gamma\delta$-lineage differentiation does not require Notch signaling in addition to the strong TCR signal to allow further differentiation into mature DN $\gamma\delta$ T cells.

As also discussed earlier, preTCR driven differentiation of human pre-β-selection thymocytes into true $CD4^+CD8\beta^+$ DP thymocytes occurs more efficiently in conditions of lower Notch activation compared to conditions of high Notch signaling activity, illustrating a clear difference in Notch signaling requirement for human compared to mouse $\alpha\beta$-lineage cells. While it was clearly illustrated in the mouse that differentiation and proliferation at this stage of T cell development are clearly linked and both Notch dependent (Kreslavsky et al. 2008), similar experiments show that both of these processes are clearly uncoupled in human. Further differentiation of pre-β-selection thymocytes into 'true' $CD4^+CD8\alpha\beta^+$ DP cells in human is not only independent on Notch signaling activity, it is also uncoupled of proliferation as illustrated through CFSE experiments (Fig. 4). Strong Notch activation in conjunction with pre-TCR signaling results in the maintenance of a DN phenotype and delayed differentiation, which is in contrast to T cell development in the mouse (Taghon et al. 2009; Van Coppernolle et al. 2012). With respect to the influence of Notch signal strength on human $\gamma\delta$ T cell development, it is currently known that maturation of immature $CD1^+TCR\gamma\delta^+$ thymocytes into mature $\gamma\delta$ T cells can occur virtually equally efficient in the presence or absence of Notch signaling (Van Coppernolle et al. 2012). However, it is not clear yet how Notch signaling activity affects the $\gamma\delta$- versus $\alpha\beta$-lineage choice in conjunction with strong TCR signaling since transduction experiments with a specific $\gamma\delta$ TCR have not been performed in human.

While the TCR signal strength model explains a lot of the experimental data that has been provided in light of this developmental choice, it is safe to predict that results of the initial TCR rearrangements in immature thymocytes will

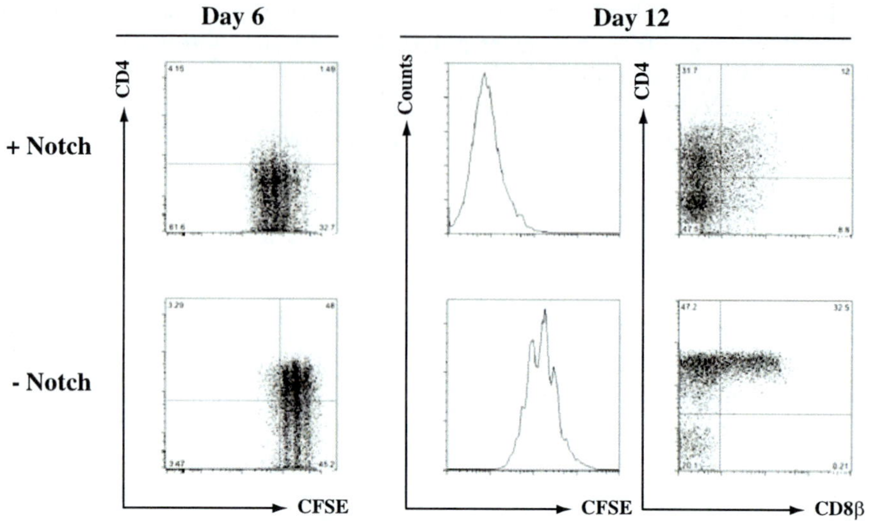

Fig. 4 Uncoupling of proliferation and differentiation during human T cell development. Human $CD34^{+}$ postnatal thymocytes were CFSE labeled following purification and cocultured on OP9-DL1 (+ Notch) or OP9-control (− Notch) stromal cells. Cultures were analyzed after 6 or 12 days and stained for CD4 and CD8β, as indicated

determine the developmental outcome of a large portion of the cells that have generated a $\gamma\delta$ or preTCR (Buer et al. 1997). Thus, the mechanisms that control these TCR rearrangements provide a first insight into the $\alpha\beta$ versus $\gamma\delta$ T-cell lineage choice. Importantly, while a critical role for Notch in *TCRB* rearrangements has been illustrated in both mouse (Wolfer et al. 2002) and human (De Smedt et al. 2005), it is unclear if TCR-γ and -δ rearrangements depend on this signaling pathway. This obscures a clear interpretation of the early effects of Notch signaling on mouse T cell development since in experiments in which Notch activity was conditionally affected, deletion was induced following the initiation of T cell specification during which rearrangements of the *TCRG* and *-D* loci are already initiated (Tanigaki et al. 2004; Radtke et al. 1999). As such, it is difficult to interpret if $\gamma\delta$ T cells in the mouse might not also depend on Notch activity early on to allow these rearrangement events. Experiments with purified DN2 subsets did reveal a clear reduction in $\gamma\delta$ T cell in the absence of Notch signaling (Ciofani et al. 2004). Mechanistically, this could be mediated through regulation of *IL7R* expression (Gonzalez-Garcia et al. 2009) since signaling through this receptor has been shown to be essential for *TCRG* rearrangements (Maki et al. 1996; Durum et al. 1998). As such, high Notch signaling activity in human could promote $\gamma\delta$ T cell development by preferentially inducing rearrangements at the gene segments of these TCRs (Ye et al. 2001).

Thus, Notch signaling might mediate the $\alpha\beta/\gamma\delta$-lineage choice at two distinct stages of T cell development. A first role may involve regulation of *TCRD*, *-G* and *-B* rearrangements in immature thymocytes prior to $\gamma\delta$- or β-selection, while a second role becomes apparent following these TCR-mediated selection processes as Notch activity synergizes with these $\gamma\delta$ and preTCR signaling events. As such, further studies will be essential to determine how different the role of Notch signaling is in these events during human and mouse T cell development.

5 Conclusion

While the current experimental data reveals both similar and alternative roles for the Notch signaling pathway during human T cell development compared to mouse, much work is still required to fully understand the precise mechanisms through which Notch signaling influences each of these early stages of this developmental process. While the role of differential Notch receptor/ligands remains to be explored, also the integration of various molecular mechanisms downstream of the Notch receptors into the global transcriptional network that drives human T cell development is still in its infancy. While much more progress has been made with respect to these issues in the mouse (Kueh and Rothenberg 2012; Radtke et al. 2010; Maillard et al. 2005; Yuan et al. 2010), we anticipate that novel technical advances, that enable, for instance, knockout and gene-reporter approaches in human ES cells, will yield more definitive insights into these processes (Timmermans et al. 2009; Galic et al. 2006; Hockemeyer et al. 2009; Hockemeyer et al. 2011). Given that recent insights suggest important regulatory differences for the Notch signaling pathway in human versus mouse T-acute lymphoblastic leukemia (Wang et al. 2011), tackling these questions is of vital importance for understanding normal and malignant developmental processes in human.

Acknowledgments The authors wish to thank Jean Plum for critical reading of the manuscript. Work on this subject by the Taghon lab is funded by a research grant from the Fund for Scientific Research Flanders (FWO), as well as by its Odysseus program. IVdW is supported by a grant from the Special Fund for Scientific Research (BOF) of the Ghent University and TT by a fellowship of the FWO.

References

Awong G, Herer E, Surh CD, Dick JE, La Motte-Mohs RN, Zuniga-Pflucker JC (2009) Characterization in vitro and engraftment potential in vivo of human progenitor T cells generated from hematopoietic stem cells. Blood 114:972–982

Beatus P, Lundkvist J, Oberg C, Lendahl U (1999) The notch 3 intracellular domain represses notch 1-mediated activation through Hairy/Enhancer of split (HES) promoters. Development 126:3925–3935

Benne C, Lelievre JD, Balbo M, Henry A, Sakano S, Levy Y (2009) Notch increases T/NK potential of human hematopoietic progenitors and inhibits B cell differentiation at a pro-B stage. Stem Cells 27:1676–1685

Blom B, Spits H (2006) Development of human lymphoid cells. Annu Rev Immunol 24:287–320
Buer J, Aifantis I, DiSanto JP, Fehling HJ, von Boehmer H (1997) Role of different T cell receptors in the development of pre-T cells. J Exp Med 185:1541–1547
Carrasco YR, Trigueros C, Ramiro AR, de Yebenes V, Toribio ML (1999) Beta-selection is associated with the onset of CD8beta chain expression on CD4(+)CD8alphaalpha(+) pre-T cells during human intrathymic development. Blood 94:3491–3498
Ciofani M, Schmitt TM, Ciofani A, Michie AM, Cuburu N, Aublin A, Maryanski JL, Zuniga-Pflucker JC (2004) Obligatory role for cooperative signaling by pre-TCR and Notch during thymocyte differentiation. J Immunol 172:5230–5239
Ciofani M, Zuniga-Pflucker JC (2005) Notch promotes survival of pre-T cells at the beta-selection checkpoint by regulating cellular metabolism. Nat Immunol 6:881–888
Ciofani M, Zuniga-Pflucker JC (2010) Determining gammadelta versus alphass T cell development. Nat Rev Immunol 10:657–663
David-Fung ES, Yui MA, Morales M, Wang H, Taghon T, Diamond RA, Rothenberg EV (2006) Progression of regulatory gene expression states in fetal and adult pro-T-cell development. Immunol Rev 209:212–236
De Smedt M, Hoebeke I, Plum J (2004) Human bone marrow CD34 + progenitor cells mature to T cells on OP9-DL1 stromal cell line without thymus microenvironment. Blood Cells Mol Dis 33:227–232
De Smedt M, Hoebeke I, Reynvoet K, Leclercq G, Plum J (2005) Different thresholds of Notch signaling bias human precursor cells toward B-, NK-, monocytic/dendritic-, or T-cell lineage in thymus microenvironment. Blood 106:3498–3506
De Smedt M, Reynvoet K, Kerre T, Taghon T, Verhasselt B, Vandekerckhove B, Leclercq G, Plum J (2002) Active form of Notch imposes T cell fate in human progenitor cells. J Immunol 169:3021–3029
De Smedt M, Taghon T, Van de Walle I, De Smet G, Leclercq G, Plum J (2007) Notch signaling induces cytoplasmic CD3 epsilon expression in human differentiating NK cells. Blood 110:2696–2703
Dik WA, Pike-Overzet K, Weerkamp F, de Ridder D, de Haas EF, Baert MR, van der Spek P, Koster EE, Reinders MJ, van Dongen JJ, Langerak AW, Staal FJ (2005) New insights on human T cell development by quantitative T cell receptor gene rearrangement studies and gene expression profiling. J Exp Med 201:1715–1723
Doerfler P, Shearman MS, Perlmutter RM (2001) Presenilin-dependent gamma-secretase activity modulates thymocyte development. Proc Natl Acad Sci USA 98:9312–9317
Dontje W, Schotte R, Cupedo T, Nagasawa M, Scheeren F, Gimeno R, Spits H, Blom B (2006) Delta-like1-induced Notch1 signaling regulates the human plasmacytoid dendritic cell versus T-cell lineage decision through control of GATA-3 and Spi-B. Blood 107:2446–2452
Doulatov S, Notta F, Eppert K, Nguyen LT, Ohashi PS, Dick JE (2010) Revised map of the human progenitor hierarchy shows the origin of macrophages and dendritic cells in early lymphoid development. Nat Immunol 11:585–593
Durum SK, Candeias S, Nakajima H, Leonard WJ, Baird AM, Berg LJ, Muegge K (1998) Interleukin 7 receptor control of T cell receptor gamma gene rearrangement: role of receptor-associated chains and locus accessibility. J Exp Med 188:2233–2241
Feyerabend TB, Terszowski G, Tietz A, Blum C, Luche H, Gossler A, Gale NW, Radtke F, Fehling HJ, Rodewald HR (2009) Deletion of Notch1 converts pro-T cells to dendritic cells and promotes thymic B cells by cell-extrinsic and cell-intrinsic mechanisms. Immunity 30:67–79
Galic Z, Kitchen SG, Kacena A, Subramanian A, Burke B, Cortado R, Zack JA (2006) T lineage differentiation from human embryonic stem cells. Proc Natl Acad Sci USA 103:11742–11747
Galy A, Travis M, Cen D, Chen B (1995) Human T, B, natural killer, and dendritic cells arise from a common bone marrow progenitor cell subset. Immunity 3:459–473
Garbe AI, Krueger A, Gounari F, Zuniga-Pflucker JC, von Boehmer H (2006) Differential synergy of Notch and T cell receptor signaling determines alphabeta versus gammadelta lineage fate. J Exp Med 203:1579–1590

Garcia-Peydro M, de Yebenes V, Toribio ML (2003) Sustained Notch1 signaling instructs the earliest human intrathymic precursors to adopt a gammadelta T-cell fate in fetal thymus organ culture. Blood 102:2444–2451

Ghisi M, Corradin A, Basso K, Frasson C, Serafin V, Mukherjee S, Mussolin L, Ruggero K, Bonanno L, Guffanti A, De Bellis G, Gerosa G, Stellin G, D'Agostino DM, Basso G, Bronte V, Indraccolo S, Amadori A, Zanovello P (2011) Modulation of microRNA expression in human T-cell development: targeting of NOTCH3 by miR-150. Blood 117:7053–7062

Gonzalez-Garcia S, Garcia-Peydro M, Martin-Gayo E, Ballestar E, Esteller M, Bornstein R, de la Pompa JL, Ferrando AA, Toribio ML (2009) CSL-MAML-dependent Notch1 signaling controls T lineage-specific IL-7R{alpha} gene expression in early human thymopoiesis and leukemia. J Exp Med 206:779–791

Gotter J, Brors B, Hergenhahn M, Kyewski B (2004) Medullary epithelial cells of the human thymus express a highly diverse selection of tissue-specific genes colocalized in chromosomal clusters. J Exp Med 199:155–166

Green MD, Chen A, Nostro MC, d'Souza SL, Schaniel C, Lemischka IR, Gouon-Evans V, Keller G, Snoeck HW (2011) Generation of anterior foregut endoderm from human embryonic and induced pluripotent stem cells. Nat Biotechnol 29:267–272

Haddad R, Guimiot F, Six E, Jourquin F, Setterblad N, Kahn E, Yagello M, Schiffer C, Andre-Schmutz I, Cavazzana-Calvo M, Gluckman JC, Delezoide AL, Pflumio F, Canque B (2006) Dynamics of thymus-colonizing cells during human development. Immunity 24:217–230

Hao QL, George AA, Zhu J, Barsky L, Zielinska E, Wang X, Price M, Ge S, Crooks GM (2008) Human intrathymic lineage commitment is marked by differential CD7 expression: identification of CD7- lympho-myeloid thymic progenitors. Blood 111:1318–1326

Hao QL, Zhu J, Price MA, Payne KJ, Barsky LW, Crooks GM (2001) Identification of a novel, human multilymphoid progenitor in cord blood. Blood 97:3683–3690

Heinzel K, Benz C, Martins VC, Haidl ID, Bleul CC (2007) Bone marrow-derived hemopoietic precursors commit to the T cell lineage only after arrival in the thymic microenvironment. J Immunol 178:858–868

Hirata H, Yoshiura S, Ohtsuka T, Bessho Y, Harada T, Yoshikawa K, Kageyama R (2002) Oscillatory expression of the bHLH factor Hes1 regulated by a negative feedback loop. Science 298:840–843

Hockemeyer D, Soldner F, Beard C, Gao Q, Mitalipova M, DeKelver RC, Katibah GE, Amora R, Boydston EA, Zeitler B, Meng X, Miller JC, Zhang L, Rebar EJ, Gregory PD, Urnov FD, Jaenisch R (2009) Efficient targeting of expressed and silent genes in human ESCs and iPSCs using zinc-finger nucleases. Nat Biotechnol 27:851–857

Hockemeyer D, Wang H, Kiani S, Lai CS, Gao Q, Cassady JP, Cost GJ, Zhang L, Santiago Y, Miller JC, Zeitler B, Cherone JM, Meng X, Hinkley SJ, Rebar EJ, Gregory PD, Urnov FD, Jaenisch R (2011) Genetic engineering of human pluripotent cells using TALE nucleases. Nat Biotechnol 29:731–734

Hoebeke I, De Smedt M, Stolz F, Pike-Overzet K, Staal FJ, Plum J, Leclercq G (2007) T-, B- and NK-lymphoid, but not myeloid cells arise from human CD34(+)CD38(-)CD7(+) common lymphoid progenitors expressing lymphoid-specific genes. L. Leukemia 21:311–319

Hosoya T, Kuroha T, Moriguchi T, Cummings D, Maillard I, Lim KC, Engel JD (2009) GATA-3 is required for early T lineage progenitor development. J Exp Med 206:2987–3000

Hozumi K, Mailhos C, Negishi N, Hirano KI, Yahata T, Ando K, Zuklys S, Hollander GA, Shima DT, Habu S (2008) Delta-like 4 is indispensable in thymic environment specific for T cell development. J Exp Med

Hozumi K, Negishi N, Suzuki D, Abe N, Sotomaru Y, Tamaoki N, Mailhos C, Ish-Horowicz D, Habu S, Owen MJ (2004) Delta-like 1 is necessary for the generation of marginal zone B cells but not T cells in vivo. Nat Immunol 5:638–644

Ikawa T, Hirose S, Masuda K, Kakugawa K, Satoh R, Shibano-Satoh A, Kominami R, Katsura Y, Kawamoto H (2010) An essential developmental checkpoint for production of the T cell lineage. Science 329:93–96

Ikawa T, Kawamoto H, Goldrath AW, Murre C (2006) E proteins and Notch signaling cooperate to promote T cell lineage specification and commitment. J Exp Med 203:1329–1342
Jaleco AC, Neves H, Hooijberg E, Gameiro P, Clode N, Haury M, Henrique D, Parreira L (2001) Differential effects of Notch ligands Delta-1 and Jagged-1 in human lymphoid differentiation. J Exp Med 194:991–1002
Joachims ML, Chain JL, Hooker SW, Knott-Craig CJ, Thompson LF (2006) Human alpha beta and gamma delta thymocyte development: TCR gene rearrangements, intracellular TCR beta expression, and gamma delta developmental potential–differences between men and mice. J Immunol 176:1543–1552
Klein F, Feldhahn N, Lee S, Wang H, Ciuffi F, von Elstermann M, Toribio ML, Sauer H, Wartenberg M, Barath VS, Kronke M, Wernet P, Rowley JD, Muschen M (2003) T lymphoid differentiation in human bone marrow. Proc Natl Acad Sci USA 100:6747–6752
Koch U, Fiorini E, Benedito R, Besseyrias V, Schuster-Gossler K, Pierres M, Manley NR, Duarte A, Macdonald HR, Radtke F (2008) Delta-like 4 is the essential, nonredundant ligand for Notch1 during thymic T cell lineage commitment. J Exp Med
Kreslavsky T, Garbe AI, Krueger A, von Boehmer H (2008) T cell receptor-instructed alphabeta versus gammadelta lineage commitment revealed by single-cell analysis. J Exp Med 205:1173–1186
Kreslavsky T, Gleimer M, Garbe AI, von Boehmer H (2010) alphabeta versus gammadelta fate choice: counting the T-cell lineages at the branch point. Immunol Rev 238:169–181
Krueger A, Garbe AI, von Boehmer H (2006) Phenotypic plasticity of T cell progenitors upon exposure to Notch ligands. J Exp Med 203:1977–1984
Kueh HY, Rothenberg EV (2012) Regulatory gene network circuits underlying T cell development from multipotent progenitors. Wiley Interdiscip Rev Syst Biol Med 4:79–102
La Motte-Mohs RN, Herer E, Zuniga-Pflucker JC (2005) Induction of T-cell development from human cord blood hematopoietic stem cells by Delta-like 1 in vitro. Blood 105:1431–1439
Lee SY, Stadanlick J, Kappes DJ, Wiest DL (2010) Towards a molecular understanding of the differential signals regulating alphabeta/gammadelta T lineage choice. Semin Immunol 22:237–246
Lefort N, Benne C, Lelievre JD, Dorival C, Balbo M, Sakano S, Coulombel L, Levy Y (2006) Short exposure to Notch ligand Delta-4 is sufficient to induce T-cell differentiation program and to increase the T cell potential of primary human CD34 + cells. Exp Hematol 34:1720–1729
Lehar SM, Dooley J, Farr AG, Bevan MJ (2005) Notch ligands Delta 1 and Jagged1 transmit distinct signals to T-cell precursors. Blood 105:1440–1447
Li K, Li Y, Wu W, Gordon WR, Chang DW, Lu M, Scoggin S, Fu T, Vien L, Histen G, Zheng J, Martin-Hollister R, Duensing T, Singh S, Blacklow SC, Yao Z, Aster JC, Zhou BB (2008) Modulation of Notch signaling by antibodies specific for the extracellular negative regulatory region of NOTCH3. J Biol Chem 283:8046–8054
Li L, Leid M, Rothenberg EV (2010a) An early T cell lineage commitment checkpoint dependent on the transcription factor Bcl11b. Science 329:89–93
Li P, Burke S, Wang J, Chen X, Ortiz M, Lee SC, Lu D, Campos L, Goulding D, Ng BL, Dougan G, Huntly B, Gottgens B, Jenkins NA, Copeland NG, Colucci F, Liu P (2010b) Reprogramming of T cells to natural killer-like cells upon Bcl11b deletion. Science 329:85–89
Magri M, Yatim A, Benne C, Balbo M, Henry A, Serraf A, Sakano S, Gazzolo L, Levy Y, Lelievre JD (2009) Notch ligands potentiate IL-7-driven proliferation and survival of human thymocyte precursors. Eur J Immunol 39:1231–1240
Maillard I, Fang T, Pear WS (2005) Regulation of lymphoid development, differentiation, and function by the Notch pathway. Annu Rev Immunol 23:945–974
Maillard I, Tu L, Sambandam A, Yashiro-Ohtani Y, Millholland J, Keeshan K, Shestova O, Xu L, Bhandoola A, Pear WS (2006) The requirement for Notch signaling at the beta-selection checkpoint in vivo is absolute and independent of the pre-T cell receptor. J Exp Med 203:2239–2245
Maki K, Sunaga S, Ikuta K (1996) The V-J recombination of T cell receptor-gamma genes is blocked in interleukin-7 receptor-deficient mice. J Exp Med 184:2423–2427

Neves H, Weerkamp F, Gomes AC, Naber BA, Gameiro P, Becker JD, Lucio P, Clode N, van Dongen JJ, Staal FJ, Parreira L (2006) Effects of Delta1 and Jagged1 on early human hematopoiesis: correlation with expression of notch signaling-related genes in CD34 + cells. Stem Cells 24:1328–1337

Palomero T, Lim WK, Odom DT, Sulis ML, Real PJ, Margolin A, Barnes KC, O'Neil J, Neuberg D, Weng AP, Aster JC, Sigaux F, Soulier J, Look AT, Young RA, Califano A, Ferrando AA (2006) NOTCH1 directly regulates c-MYC and activates a feed-forward-loop transcriptional network promoting leukemic cell growth. Proc Natl Acad Sci USA 103:18261–18266

Petrie HT, Zuniga-Pflucker JC (2007) Zoned out: functional mapping of stromal signaling microenvironments in the thymus. Annu Rev Immunol 25:649–679

Plum J, De Smedt M, Leclercq G, Taghon T, Kerre T, Vandekerckhove B (2008) Human intrathymic development: a selective approach. Semin Immunopathol 30:411–423

Radtke F, Fasnacht N, Macdonald HR (2010) Notch signaling in the immune system. Immunity 32:14–27

Radtke F, Wilson A, Stark G, Bauer M, van Meerwijk J, Macdonald HR, Aguet M (1999) Deficient T cell fate specification in mice with an induced inactivation of Notch1. Immunity 10:547–558

Reizis B, Leder P (2002) Direct induction of T lymphocyte-specific gene expression by the mammalian Notch signaling pathway. Genes Dev 16:295–300

Robey E, Chang D, Itano A, Cado D, Alexander H, Lans D, Weinmaster G, Salmon P (1996) An activated form of Notch influences the choice between CD4 and CD8 T cell lineages. Cell 87:483–492

Sambandam A, Maillard I, Zediak VP, Xu L, Gerstein RM, Aster JC, Pear WS, Bhandoola A (2005) Notch signaling controls the generation and differentiation of early T lineage progenitors. Nat Immunol 6:663–670

Schmitt TM, Ciofani M, Petrie HT, Zuniga-Pflucker JC (2004) Maintenance of T cell specification and differentiation requires recurrent notch receptor-ligand interactions. J Exp Med 200:469–479

Shi J, Fallahi M, Luo JL, Petrie HT (2011) Nonoverlapping functions for Notch1 and Notch3 during murine steady-state thymic lymphopoiesis. Blood 118:2511–2519

Six EM, Bonhomme D, Monteiro M, Beldjord K, Jurkowska M, Cordier-Garcia C, Garrigue A, Dal CL, Rocha B, Fischer A, Cavazzana-Calvo M, Andre-Schmutz I (2007) A human postnatal lymphoid progenitor capable of circulating and seeding the thymus. J Exp Med 204:3085–3093

Suliman S, Tan J, Xu K, Kousis PC, Kowalski PE, Chang G, Egan SE, Guidos C (2011) Notch3 is dispensable for thymocyte beta-selection and Notch1-induced T cell leukemogenesis. PLoS One 6:e24937

Taghon T, De Smedt M, Stolz F, Cnockaert M, Plum J, Leclercq G (2001) Enforced expression of GATA-3 severely reduces human thymic cellularity. J Immunol 167:4468–4475

Taghon T, Rothenberg EV (2008) Molecular mechanisms that control mouse and human TCR-alphabeta and TCR-gammadelta T cell development. Semin Immunopathol 30:383–398

Taghon T, Stolz F, De Smedt M, Cnockaert M, Verhasselt B, Plum J, Leclercq G (2002) HOX-A10 regulates hematopoietic lineage commitment: evidence for a monocyte-specific transcription factor. Blood 99:1197–1204

Taghon T, Van de Walle I, De Smet G, De Smedt M, Leclercq G, Vandekerckhove B, Plum J (2009) Notch signaling is required for proliferation but not for differentiation at a well-defined beta-selection checkpoint during human T-cell development. Blood 113:3254–3263

Taghon T, Yui MA, Pant R, Diamond RA, Rothenberg EV (2006) Developmental and molecular characterization of emerging beta- and gammadelta-selected pre-T cells in the adult mouse thymus. Immunity 24:53–64

Taghon T, Yui MA, Rothenberg EV (2007) Mast cell lineage diversion of T lineage precursors by the essential T cell transcription factor GATA-3. Nat Immunol 8:845–855

Taghon TN, David ES, Zuniga-Pflucker JC, Rothenberg EV (2005) Delayed, asynchronous, and reversible T-lineage specification induced by Notch/Delta signaling. Genes Dev 19:965–978

Tanigaki K, Tsuji M, Yamamoto N, Han H, Tsukada J, Inoue H, Kubo M, Honjo T (2004) Regulation of alphabeta/gammadelta T cell lineage commitment and peripheral T cell responses by Notch/RBP-J signaling. Immunity 20:611–622

Timmermans F, Velghe I, Vanwalleghem L, De Smedt M, Van Coppernolle S, Taghon T, Moore HD, Leclercq G, Langerak AW, Kerre T, Plum J, Vandekerckhove B (2009) Generation of T cells from human embryonic stem cell-derived hematopoietic zones. J Immunol 182:6879–6888

Ting CN, Olson MC, Barton KP, Leiden JM (1996) Transcription factor GATA-3 is required for development of the T-cell lineage. Nature 384:474–478

Tydell CC, vid-Fung ES, Moore JE, Rowen L, Taghon T, Rothenberg EV (2007) Molecular dissection of prethymic progenitor entry into the T lymphocyte developmental pathway. J Immunol 179:421–438

Van Coppernolle S, Vanhee S, Verstichel G, Snauwaert S, van der Spek A, Velghe I, Sinnesael M, Heemskerk MH, Taghon T, Leclercq G, Plum J, Langerak AW, Kerre T, Vandekerckhove B (2012) Notch induces human T-cell receptor gammadelta + thymocytes to differentiate along a parallel, highly proliferative and bipotent CD4 CD8 double-positive pathway. Leukemia 26:127–138

Van de Walle I, De Smet G, Gartner M, De Smedt M, Waegemans E, Vandekerckhove B, Leclercq G, Plum J, Aster JC, Bernstein ID, Guidos CJ, Kyewski B, Taghon T (2011) Jagged2 acts as a Delta-like Notch ligand during early hematopoietic cell fate decisions. Blood 117:4449–4459

Van de Walle I, De Smet G, De Smedt M, Vandekerckhove B, Leclercq G, Plum J, Taghon T (2009) An early decrease in Notch activation is required for human TCR-alphabeta lineage differentiation at the expense of TCR-gammadelta T cells. Blood 113:2988–2998

Vandekerckhove B, Vanhee S, Van Coppernolle S, Snauwaert S, Velghe I, Taghon T, Leclercq G, Kerre T, Plum J (2011) In vitro generation of immune cells from pluripotent stem cells. Front Biosci 16:1488–1504

Verbeek S, Izon D, Hofhuis F, Robanus-Maandag E, te Riele H, van de Wetering M, Oosterwegel M, Wilson A, Macdonald HR, Clevers H (1995) An HMG-box-containing T-cell factor required for thymocyte differentiation. Nature 374:70–74

Wang H, Zou J, Zhao B, Johannsen E, Ashworth T, Wong H, Pear WS, Schug J, Blacklow SC, Arnett KL, Bernstein BE, Kieff E, Aster JC (2011) Genome-wide analysis reveals conserved and divergent features of Notch1/RBPJ binding in human and murine T-lymphoblastic leukemia cells. Proc Natl Acad Sci USA 108:14908–14913

Weber BN, Chi AW, Chavez A, Yashiro-Ohtani Y, Yang Q, Shestova O, Bhandoola A (2011) A critical role for TCF-1 in T-lineage specification and differentiation. Nature 476:63–68

Weerkamp F, Baert MR, Brugman MH, Dik WA, de Haas EF, Visser TP, de Groot CJ, Wagemaker G, van Dongen JJ, Staal FJ (2006a) Human thymus contains multipotent progenitors with T/B lymphoid, myeloid, and erythroid lineage potential. Blood 107:3131–3137

Weerkamp F, van Dongen JJ, Staal FJ (2006b) Notch and Wnt signaling in T-lymphocyte development and acute lymphoblastic leukemia. Leukemia 20:1197–1205

Weng AP, Millholland JM, Yashiro-Ohtani Y, Arcangeli ML, Lau A, Wai C, Del Bianco C, Rodriguez CG, Sai H, Tobias J, Li Y, Wolfe MS, Shachaf C, Felsher D, Blacklow SC, Pear WS, Aster JC (2006) c-Myc is an important direct target of Notch1 in T-cell acute lymphoblastic leukemia/lymphoma. Genes Dev 20:2096–2109

Wilson A, Macdonald HR, Radtke F (2001) Notch 1-deficient common lymphoid precursors adopt a B cell fate in the thymus. J Exp Med 194:1003–1012

Wolfer A, Wilson A, Nemir M, Macdonald HR, Radtke F (2002) Inactivation of Notch1 impairs VDJbeta rearrangement and allows pre-TCR-independent survival of early alpha beta Lineage Thymocytes. Immunity 16:869–879

Wu Y, Cain-Hom C, Choy L, Hagenbeek TJ, de Leon GP, Chen Y, Finkle D, Venook R, Wu X, Ridgway J, Schahin-Reed D, Dow GJ, Shelton A, Stawicki S, Watts RJ, Zhang J, Choy R, Howard P, Kadyk L, Yan M, Zha J, Callahan CA, Hymowitz SG, Siebel CW (2010) Therapeutic antibody targeting of individual Notch receptors. Nature 464:1052–1057

Ye SK, Agata Y, Lee HC, Kurooka H, Kitamura T, Shimizu A, Honjo T, Ikuta K (2001) The IL-7 receptor controls the accessibility of the TCRgamma locus by Stat5 and histone acetylation. Immunity 15:813–823

Yuan JS, Kousis PC, Suliman S, Visan I, Guidos CJ (2010) Functions of notch signaling in the immune system: consensus and controversies. Annu Rev Immunol 28:343–365

Yui MA, Feng N, Rothenberg EV (2010) Fine-scale staging of T cell lineage commitment in adult mouse thymus. J Immunol 185:284–293

Canonical and Non-Canonical Notch Signaling in CD4$^+$ T Cells

Lisa M. Minter and Barbara A. Osborne

Abstract For T cells to become fully activated, they must integrate a myriad of signals, both extrinsic and intrinsic. External stimuli accrued through various cell surface receptors are transduced and amplified through a coordinated circuitry of signaling cascades that ultimately result in the transcription of new genes. Along the way, extracellular and intracellular signaling components function to impart a fully activated state. Evidence is accumulating to show that the Notch family of cell surface receptors, long known to function as transcriptional regulators through their interactions with the canonical nuclear binding protein CSL/RBP-J, may also be playing an as-yet-unappreciated role in T cell activation by virtue of its signaling via non-canonical as well as nonnuclear mechanisms. In this review we will discuss these and other better-known means by which Notch signaling influences T cell responses.

Contents

L. M. Minter · B. A. Osborne (✉)
Department of Veterinary and Animal Sciences,
Program in Molecular and Cellular Biology,
University of Massachusetts, Amherst, MA 01003, USA
e-mail: osborne@vasci.umass.edu

L. M. Minter
e-mail: lminter@vasci.umass.edu

Current Topics in Microbiology and Immunology (2012) 360: 99–114
DOI: 10.1007/82_2012_233

Published Online: 15 June 2012

1 Notch Signaling in CD4$^+$ T Cells

In activated T cells, canonical Notch signaling is initiated when Notch receptors (Notch1-4) on receptor-bearing cells interact with cognate ligand from the Delta-like (Dll1, 3, 4) or Jagged (Jag1, 2) families expressed on nearby ligand-bearing cells. This physical binding is thought to generate mechanical force as the extracellular domain of the Notch receptor is pulled outward away from the surface of the receptor-bearing cell and toward the surface of the ligand-bearing cell, into which it is eventually endocytosed (Klueg and Muskavitch 1999). Surface-bound ADAM10 facilitates Notch receptor cleavage from the cell surface by severing the receptor at the S2 cleavage site, close to the exterior of the plasma membrane (van Tetering et al. 2009). Removing the extracellular domain of the receptor allows its remaining transmembrane-bound form to be cleaved in the early endosome by the enzymatic actions of the gamma-secretase complex (Saxena et al. 2001). Once liberated from the membrane, the signaling-competent intracellular domain of the Notch receptor (NICD) is imported into the nucleus where it regulates the expression of target genes. During canonical Notch signaling, NICD is loaded onto its nuclear binding partner, CSL (CBF1-Suppressor of Hairless-Lag1/RBP-j) where its presence in a transcriptional complex converts the actions of CSL from a repressor to an activator of transcription (Hsieh et al. 1996). During this transition, classical corepressors such as CIR (CBF1-interacting coRepressor) and SMRT (Silencing Mediator of Retinoid and Thyroid hormone receptors) which interact with SKIP (Ski-interacting protein) at a site distant from NICD binding, are replaced in the DNA-bound complex with coactivators, including Mastermind-like (MAML; Wu et al. 2000) and p300 (Hansson et al. 2009).

Notch receptors have been shown to colocalize with CD4 (Benson et al. 2005) as well as with p56lck (Sade et al. 2004), following T cell activation, data which suggest it is highly integrated with key molecular events occurring immediately downstream of signaling through the T cell receptor. Additionally, there are differences in temporal expression of Notch receptors, with Notch1 upregulated as early as 4 h after stimulation with αCD3ε and showing sustained expression up to 48 h after T cell activation. In contrast, upregulation of Notch2 lags behind that of Notch1, with its expression peaking at 24 h and returning to basal levels within 48 h (Fiorini et al. 2009). Early and concurrent work from our lab and the Pear and Turka labs defined roles for Notch in mediating IL-2 and IFNγ production (Palaga et al. 2003), as well as expression of CD25, the α chain of the high-affinity IL-2 receptor (Adler et al. 2003). Both groups used an inhibitor of γ-secretase to prevent release of the transcriptionally competent

NICD, and the resulting, complementary data pointed to a role for Notch in lowering the threshold for T cell activation by modulating the strength of signal. Since these initial experiments were performed, several groups have examined the effects of specific ligands on T cell activation, using markers such as CD25 and CD69 as indicators of activation. What has emerged is an apparent hierarchy of signal strength imparted by unique ligands, with Dll4 providing a T cell activation profile that looks identical to T cells stimulated with αCD3 in the absence of ligand. This is followed, in order of decreasing activation potential, by Dll1 and Jagged1 (Rutz et al. 2005). Thus, it does appear that signaling through Notch receptors affects the strength of signal during T cell activation, and it is the specificity of ligand that seems to dictate this response.

These results provided a new lens with which to examine a previously identified phenomenon whereby the strength of T cell activation influenced the differentiation of the mature T cells into helper type-1 or type-2 subsets (Tao et al. 1997). One could extrapolate the findings summarized above, to presume that T cells whose Notch receptors encounter Dll4 or Dll1 ligands during activation would adopt a Th1 cell fate while those interacting with Jagged1 would preferentially assume a Th2 phenotype. In fact, this hypothesis appears to be well supported by the experimental data. In their seminal paper, Flavell and colleagues identified a role for Dll family ligands in driving Th1 differentiation while Jag ligands promoted a Th2 cell fate (Amsen et al. 2004). Numerous in vivo experiments have upheld these observations, primarily in the context of various autoimmune models. Using a combination of approaches, including delivering neutralizing antibodies or specific ligands in the form of Fc-fusion peptides, the pathology of classically defined Th1/Th17- or Th2-mediated conditions has been successfully manipulated to demonstrate the overarching influence of ligands on T cell differentiation in vivo (Elyaman et al. 2007; Takeichi et al. 2010; Bassil et al. 2011; Reynolds et al. 2011). Thus, two modes whereby Notch signaling influences T cell differentiation (strength of signal and specificity of ligand) appear to overlap with the same unifying outcome: ligand specificity dictates strength of signal which influences cell fate decisions (Fig. 1).

2 Activation of Notch in CD4⁺ T Cells

The biological outcomes downstream of Notch signaling are complex and pleiotropic and many questions remain unanswered as to how Notch mediates its many diverse and, at times, contradictory effects. One such question is "How exactly does Notch become activated in T cells?" While it is apparent that T cells exist in an environment rich in cell types, such as antigen presenting cells, known to express Notch ligands, it also is well documented that signals through the T cell receptor (TCR) are sufficient to induce γ-secretase-mediated cleavage of Notch (Adler et al. 2003; Palaga et al. 2003). $CD4^+$ T cells express active NICD within 3–4 h following engagement of the TCR. Either αCD3 or αCD3/CD28 treatment of purified $CD4^+$ T cells leads to the production of NICD. While activated $CD4^+$ T cells can express Notch ligands, they are not found in abundance on resting T cells, raising the possibility that Notch may be activated in the absence of ligand (Osborne unpublished results).

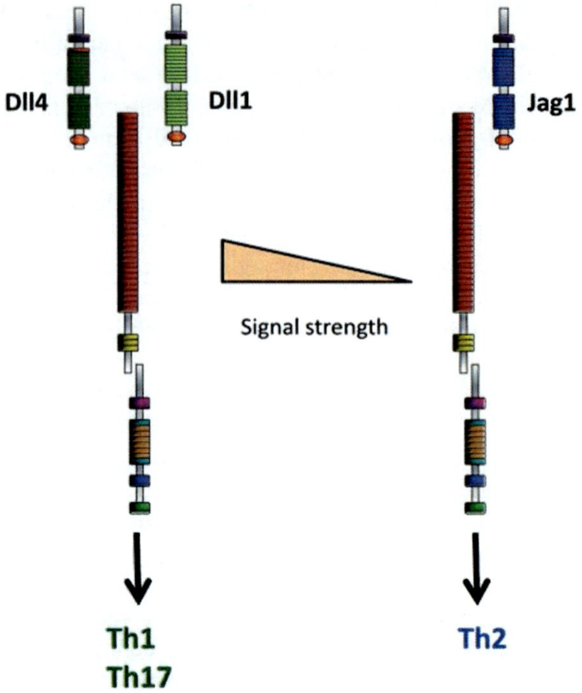

Fig. 1 Notch ligand specificity may have different biological outcomes. Data from several laboratories suggest that Dll4 and Dll1 induction of Notch signaling may drive a Th1 or Th17 response while Jagged/Notch interactions may result in a Th2 response (Amsen et al. 2004, 2007). Additionally, as indicated above, it is suggested in the data that there is hierarchy of signal strength imparted by ligands, with Dll4 providing the strongest signal and Jagged1, the weakest (Rutz et al. 2005)

Although this issue has not yet been resolved in T cells, recent data in *Drosophila* (Hori et al. 2011) demonstrate that Notch signaling can be initiated in the absence of ligand. It is well known that endocytic trafficking is an important mechanism used to regulate the level of Notch signaling. Notch is removed from the cell surface by endocytosis and either recycled back to the cell surface or sent to a multivesicular body where it undergoes lysosomal degradation. The new data from Hori and colleagues reveal a novel mechanism for Notch activation whereby, following endocytosis, the intact Notch receptor can be diverted in the multivesicular body to a signaling pathway where the receptor encounters γ-secretase and undergoes γ-secretase mediated cleavage. These data coupled with other recent examples of ligand-independent cleavage (Mukherjee et al. 2011), demonstrate that Notch can be activated independently of ligand; however, it remains to be determined whether these same pathways play a role in Notch activation in T lymphocytes.

3 The Role of Notch in CD4$^+$ T Cell Differentiation

Notch was first described in *Drosophila* as a gene that regulated cell fate decisions in specific cell lineages. Following these seminal observations in invertebrates, the role of Notch in directing cell fate in mammalian systems was examained. Several groups, including our own, have suggested that Notch is important for determining

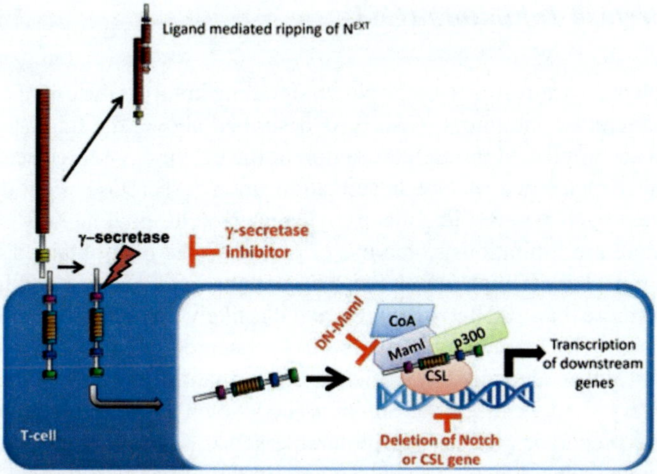

Fig. 2 Notch signaling may be inhibited by several distinct methods. Inhibition of γ-secretase blocks the release of membrane-bound $Notch_{TM}$ and thus blocks the formation of an activation complex coupled with several coactivator proteins (CoA) including Mastermind-like (MAML) and p300. Dominant-negative (DN) MAML blocks the association of this coactivator complex. Additionally, deletion of Notch genes blocks the formation of an activation complex.

mature $CD4^+$ T cell fate decisions. As mentioned above, Flavell and colleagues showed that Notch signaling, depending upon which ligand activated Notch, could direct a $CD4^+$ T cell to assume either a Th1 or a Th2 cell fate (Amsen et al. 2004, 2007). These data were followed by several conflicting reports that suggested Notch was required for and regulated Th1 (Minter et al. 2005) and not Th2 cell fate decisions or was required for Th2 and not for Th1 cell differentiation (Fang et al. 2007). More recently, Notch also has been implicated in Th17 cell fate decisions (Keerthivasan et al. 2011). It is not readily apparent how these discordant results may be resolved; however, a deeper understanding of the experimental systems used by different laboratories may provide resolution of the conflicting data and these are discussed below.

4 Different Approaches to Inhibiting Notch may Produce Conflicting Data

Much of the data discussed above derives from a wide variety of experimental systems including both in vivo as well as in vitro systems. Equally important in our consideration of what the data tell us are the methods used to block Notch signaling. Below we discuss the various methods employed to block Notch signaling and explore both the sometimes limited specificity of these methods as well as alternative interpretations of data obtained using the method in question (outlined in Fig. 2).

4.1 γ-Secretase Inhibitors (GSI)

The approach most often used to block Notch signaling involves the use of one of many available γ-Secretase Inhibitors (GSIs). As described above, the final step in Notch activation is accomplished through the action of the enzyme, γ-secretase. Because of the complete dependence of Notch activation upon γ-secretase, inhibitors of this enzyme have proved a convenient means by which Notch signaling may be inhibited. However, there are limitations to the use of GSIs and the data obtained using these compounds must be interpreted with these limitations in mind. First and foremost, γ-secretase has more than 100 known targets and it is likely that the list of substrates will continue to grow (Haapasalo and Kovacs 2011). γ-secretase has no known sequence specificity but rather recognizes a particular conformation. Transmembrane proteins with either very short extracellular domains (i.e., $Notch_{TM}$) or intracellular domains (i.e., amyloid precursor protein) are preferable substrates of this enzyme complex. The best known substrates of γ-secretase are amyloid precursor protein (APP) and Notch. However, many immunologically revelant proteins also are potential substrates of γ-secretase. In particular and pertinent to the studies covered in this review, CD43, CD44, IL-1R, and IL-6, are proteins potentially important in T cell responses, either directly or indirectly, and all are thought to be substrates of γ-secretase (Elzinga et al. 2009; Seo and Ziltener 2009; Wongchana and Palaga 2012). Therefore, interpretation of data using GSIs is complicated and investigations of Notch signaling should not be solely based on the use of GSI. Equally important for our interpretation of data using GSIs, is the recognition that not all GSIs function identically in all tissues. Unpublished data from our group and from our collaborators (Golde, Miele, Osborne, unpublished observations), demonstrated quite clearly that some GSIs are effective inhibitors of Notch signaling in breast cancer cells but are not at all effective in inhibiting Notch signaling in lymphocytes. Additionally, some GSIs are quite effective at inhibiting the cleavage of amyloid precursor protein but are less effective at blocking Notch activation. These observations led us to recognize that not all GSIs are the same and care must be taken to first characterize a particular GSI and how it acts to inhibit its substrate in the cell type under investigation. Therefore, it is clear that the use of GSI must be accompanied by other distinct experimental approaches that block Notch signaling to confirm the observations obtained with GSI.

Lastly, it must be recognized that when GSIs are used to inhibit Notch signaling, the mechanism of action is to block the production of the intracellular NICD fragment of Notch and hence, block nuclear Notch signaling. Notch protein still remains and the transmembrane portion (N_{TM}) remains intact and exposed to the cytosol. Many investigators have found that cytosolic Notch can have an effect on intracellular signaling (see below for further discussion of this possibility) and it is entirely possible that the cytosolic domain of Notch may function independently of its liberation from the plasma membrane. In such a situation, GSI would not have any effect on this unconventional nonnuclear form of Notch signaling, whereas genetic deletion of the Notch gene (and hence protein) would block such a function and reveal a role for Notch beyond the reach of GSIs.

4.2 *Mastermind as a Mechanism to Control Notch Signaling*

Another, broad approach to blocking Notch signaling is the use of a dominant negative form of Mastermind or the mammalian Mastermind-like protein (MAML). As is mentioned above, the canonical Notch signaling pathway involves the entry of NICD into the nucleus where it binds CSL and displaces corepressor complexes. NICD then recruits coactivator proteins such as p300 and MAML thus turning the CSL repressor complex into an activator complex. Truncated forms of MAML, lacking the activation domain, are known to function as dominant negatives and block transcription. Since canonical Notch signaling absolutely requires MAML, DN-MAML has been used as a mechanism to block this signaling pathway. Although the use of DN-MAML appears to be a perfectly reasonable approach to inhibit Notch signaling, recent evidence from several laboratories highlight the lack of specificity of this approach. Although DN-MAML does indeed inhibit transcription from promoters that require the MAML activator protein, it also acts as a coactivator for many other transcription pathways. There are several reports that MAML also activates p53, beta-catenin, hedgehog, NF-κB, and MEF2C transcriptional complexes and this list appears to be growing (Wu et al. 2000; Shen et al. 2006; Zhao et al. 2007; McElhinny et al. 2008; Jin et al. 2010; Vied and Kalderon 2009). Therefore, although expression of DN-MAML blocks canonical Notch signaling, it also will block transcription from the above-mentioned complexes. This complication makes it difficult to determine whether the biological effects of DN-MAML expression are exclusively due to loss of canonical Notch signaling or perhaps reflect the inhibition of additional signaling pathways.

4.3 *Experiments Involving Genetic Deletion of Notch or Notch-Related Genes Require Careful Evaluation*

The "gold standard" for determining the function of a particular protein is to genetically delete or "knock-out" the gene encoding the protein under investigation. This approach, while extremely informative, also has limitations. Some genes, such as Notch1 and Notch2, are embryonic lethal and hence, cannot be deleted in the germline (Swiatek et al. 1994; Hamada et al. 1999). Therefore, the approach for genetically deleting Notch in the immune system has been to target deletion using "floxed" alleles of the Notch gene of interest coupled with expression of Cre recombinase in the tissue under investigation. This approach, while quite rational, also has drawbacks. The most obvious limitation is that deletion of one Notch gene may be compensated for by the expression of another Notch family member. This is highlighted in a recent paper by Tacchini-Cottier and colleagues wherein they examined the response of animals challenged with *Leishmania major,* a parasite that requires a Th1 response for clearance

(Auderset et al. 2012). When this group examined mice carrying floxed alleles of Notch1 crossed to CD4-Cre mice, they noted these animals produced the Th1 cytokine, IFN-γ and cleared the infection quite effectively. A similar situation was observed when floxed Notch2 x CD4-Cre animals were infected. However when this group generated animals carrying deletions both of Notch1 and Notch2, those animals were incapable of clearing the infection and no IFN-γ was produced. Clearly, this is an instance when deletion of a single Notch family member can be misleading. Several groups have produced evidence that various Notch family members are expressed in either resting or activated peripheral T cells (reviewed in Palaga and Minter 2012). These observations, coupled with the data from the Tacchini-Cottier group demonstrate quite clearly that negative data from deletion of a single Notch gene may prove uninformative if no overt phenotype is present.

Interpretation can also be difficult when comparing studies that use conditional deletion of genes to study function. Conditional deletion relies on tissue- or temporal-specific activation of Cre recombinase. Many studies of Notch in T cells have employed either lck-Cre or more commonly CD4-Cre, each of which delete at different stages of T cell development in the thymus. Therefore, the peripheral T cells that arise in these animals have completed thymic development in the absence of the floxed Notch gene of interest. This may, however, induce the expression of another Notch gene to compensate for the loss of the deleted gene. This compensatory expression of a related Notch family member may well influence peripheral T cell function.

More recently, several labs have chosen to delete floxed alleles of Notch family members using the Mx-Cre system. Utilizing this approach allows temporal deletion of the floxed gene in mature T cells, thus allowing the T cell to fully develop in the presence of all relevant Notch family members. However, the Mx-Cre system also is not without experimental caveats. The induction of the Mx promoter requires that animals receive 3–4 injections of poly IC. Administration of this compound mimics a viral infection which, in turn, induces the expression of α/β interferons that bind to and activate the Mx promoter resulting in production of Cre recombinase. Since the poly IC injections also induce an inflammatory response, animals which express Cre recombinase using this method need to be rested for 4–5 weeks to recover from the effects of the inflammatory response. This method also requires careful controls whereby the effects on the system of poly-IC alone may be assessed.

Another possible approach for deleting a floxed gene is to introduce Cre recombinase protein directly into cells carrying floxed genes of interest. In a collaboration with our colleagues (Tew et al. unpublished observations) we have found that using a synthetic polymer based on HIV-TAT, Cre recombinase protein can be directly delivered ex vivo to isolated cells to induce deletion of floxed genes in culture. Such an approach may hold great promise for efficient in vitro deletion of floxed genes of interest.

In summary, genetic deletion of Notch genes has been very informative but there are caveats when interpreting the data from knockout studies. First of all, whether or not another Notch family member has compensated for the loss of the deleted gene must be considered. Temporal deletion by Mx-Cre, which has the advantage of allowing the immune system to mature with the full complement of genes present, induces an inflammatory response. Possible ways to temporally delete in a less disruptive fashion may be to use Tet-on/Tet-off induction of Cre recombinase or to deliver Cre recombinase using using synthetic polymers; however, these approaches have yet to be established and fully validated.

5 Canonical Versus Non-Canonical Notch Signaling

As described above, canonical Notch signaling is initiated by γ-secretase-mediated release of the intracellular form of Notch commonly referred to as NICD. Following its translocation to the nucleus, NICD associates with CSL to convert it from a repressor into an activation complex. This well-documented canonical Notch signaling pathway is conserved over evolutionary time and acts both in invertebrates as well as vertebrates to impart a wide variety of cell fate decisions.

In addition to the well-known canonical Notch signaling pathway, there are an increasing number of reports of non-canonical signaling, some of which appear to play a role in peripheral T cell function. Before we consider whether non-canonical Notch signaling is important in T cells, we will briefly define the term non-canonical and review the literature describing non-canonical Notch signaling in other systems.

The term non-canonical Notch signaling was originally coined to describe signaling events that are Notch dependent but do not rely on CSL. It was first observed in *Drosophila* using a genetic approach that examined the precise requirement of various components of the Notch signaling pathway. More recently, as our understanding of non-canonical Notch signaling has been refined, it is apparent that, at least in some instances, non-canonical Notch signaling may occur in the cytosol. We refer to this as non-canonical, nonnuclear Notch signaling.

In addition to non-canonical, nonnuclear Notch signaling there is evidence for non-canonical nuclear Notch signaling. Data from our lab have shown that Notch may likely function in a non-canonical fashion in the nucleus. Using chromatin immunoprecipitation (ChIP) to analyze the promoters of several genes important in peripheral T cell function, we found Notch associated with the NF-κB proteins, p50, or c-rel (Shin et al. 2006; Cho et al. 2009). These data suggest that Notch may regulate transcription independent of CSL.

The notion that non-canonical Notch signaling exists was hinted at in the late 1990s. In *Drosophila*, a Su(H)-independent Notch activity was implicated in dorsal closure during embryonic development (please note: Su(H) is *Drosophila* CSL) (Zecchini et al. 1999). Additionally, Giniger (1998) described a non-canonical Notch signaling pathway required for axon guidance in *Drosophila*. In vertebrates, data from Shawber et al. (1996) provide compelling evidence that myoblast differentiation occurs via non-canonical Notch signaling. The details of these non-canonical signaling pathways did not indicate whether non-canonical Notch functions occur in the cytosol or nucleus. Thus, during the past 15 years, the literature has provided ample evidence of non-canonical Notch signaling both in invertebrates as well as vertebrates.

More recently, non-canonical, cytosolic functions have been ascribed to Notch. Giniger and colleagues reexamined the issue of axon guidance in *Drosophila* and found Notch regulation of axon guidance requires interaction with *Abl*, a cytosolic tyrosine kinase (Le Gall et al. 2008). Their data provide persuasive genetic and biochemical evidence that Notch requirements for cell fate specification is completely dependent on Su(H) while axon guidance requires, instead, cytosolic interaction with Abl kinase. Thus, using genetic approaches, the Giniger group provides compelling evidence for at least two Notch signaling pathways; one being the canonical, CSL-dependent pathway and the other, a non-canonical, nonnuclear pathway.

Callahan and colleagues (Raafat et al. 2009), investigating the role of Notch signaling in mammary gland development and tumorigenesis, recently describe a clear role for canonical Notch signaling in normal mammary gland development while mammary tumor development is independent of CSL interaction and hence can be attributed to non-canonical Notch signaling. Similarly, Liu et al. (2009) report that epithelial transformation by the transcription factor, KLF4, requires Notch1 but does not require CSL and therefore occurs through non-canonical Notch signaling. Whether Notch exerts non-canonical effects in the cytosol or within the nucleus is not known; however, both of these recent reports clearly place non-canonical Notch signaling in a pathway leading to tumorigenesis.

McKay and colleagues (Androutsellis-Theotokis et al. 2006) also demonstrated a cytosolic role for Notch in the survival of neural stem cells as well as embryonic stem cells. Although they provide evidence that cytosolic Notch somehow activates the PI3 K/mTOR pathway, the mechanistic details of this effect of Notch await further investigation.

5.1 Does Non-Canonical Notch Signaling Play a Role in Peripheral T Cell Function?

The earliest evidence that non-canonical Notch signaling is important in peripheral T cell function comes from the work of the Honjo lab (Tanigaki et al. 2004 Immunity) where these investigators conditionally deleted RBP-J (another name for CSL) from T cells using CD4-Cre. The data obtained showed that IgG1 and IgE immunoglobulin levels from these animals are dramatically reduced. They then investigated the ability of these animals to produce Th1 cells and found that animals lacking CSL/RBP-J preferentially produced Th1 cells both under in vivo as well as in vitro conditions. The authors concluded that, under normal circumstances, Notch signaling represses Th1 differentiation, therefore, in the absence of CSL/RBP-J, Th1 cells are preferentially produced. However, more recent evidence from many labs (see above citations) clearly indicates that Notch does indeed play a required role in the differentiation of Th1 cells. In light of these more recent data, we suggest that another interpretation of the data from Tanigaki et al. may be that Th1 responses are regulated through non-canonical Notch signaling. A recent paper by Tacchini-Cottier, demonstrating that the Th1 response to *Leishmania* requires both Notch1 and Notch2 signaling but occurs independently of CSL/RBP-J supports the interpretation that Th1 responses require non-canonical Notch signaling (Auderset et al. 2012).

Notch has long been known to regulate apoptosis. Data from our lab over a decade ago (Jehn et al. 1999) described a role for Notch in preventing Nur77-dependent apoptosis in T cells. More recently, the pioneering work conducted by the Sarin laboratory, has enhanced our understanding of the mechanism by which Notch inhibits apoptosis. Sarin and collaborators, examining the antiapoptotic effects ascribed to Notch, demonstrated quite clearly that activation of Notch results in activation of mTOR and Akt and these effects may be localized to the cytosol (Sade et al. 2004; Perumalsamy et al. 2009, 2010). Through the use of constructs that direct the expression of Notch either to the cytosol or nucleus, the Sarin lab demonstrated that Notch localized to nucleus (N^{IC}-NLS) does not protect cells from apoptosis while Notch designed to be retained in the cytosol (N^{IC}-NES) is protective. Additionally, this group has shown that Notch-mediated activation of Akt results in maintenance of mitochondrial integrity and connectivity (Perumalsamy et al. 2010) and requires the mitochondrial associated proteins, mitofusin1 and mitofusin2. While it remains unclear how Notch activates Akt and how this involves mitofusins, these data strongly suggest that the role Notch plays in regulating apoptosis is both non-canonical and nonnuclear.

As mentioned above, we have preliminary data that support a non-canonical nuclear Notch function. Data from our lab have identified NF-κB subunits as potential nuclear interacting partners of intracellular Notch. Complexes of Notch and NF-κB1 (p50), as well as intracellular Notch and c-Rel can be immunoprecipitated bound to DNA from the promoter region of the *IFNγ* gene (Shin et al. 2006; Cho et al. 2009) indicating there may be other nuclear partners, in addition to CSL, through which intracellular Notch regulates target genes. A close examination of NF-κB- and CSL-consensus binding sites reveals an interesting finding. While not all CSL-binding motifs are subsets of a larger NF-κB response element, NF-κB consensus sites can incorporate a nested CSL binding site (Vales and Friedl 2002; Mann et al. 2002; Lee et al. 2000). In another study (Cho et al. 2009), we observed that in $CD8^+$ T cells, Notch signaling regulated expression of the *EOMES*, *perforin*, and *granzyme B* gene promoters. Interestingly, using ChIP analysis of these promoters, we identified Notch in a complex with NF-κB and treatment with GSI blocks this interaction (Fig. 3). In contrast, although CSL was found bound to all three promoters, GSI had little to no influence on CSL binding implying that CSL binding is constitutive and independent of Notch cleavage. Although we have yet to perform such experiments, we would predict that Notch/NF-κB complexes will be found on these promoters in CSL/RBP-J deficient $CD8^+$ T cells.

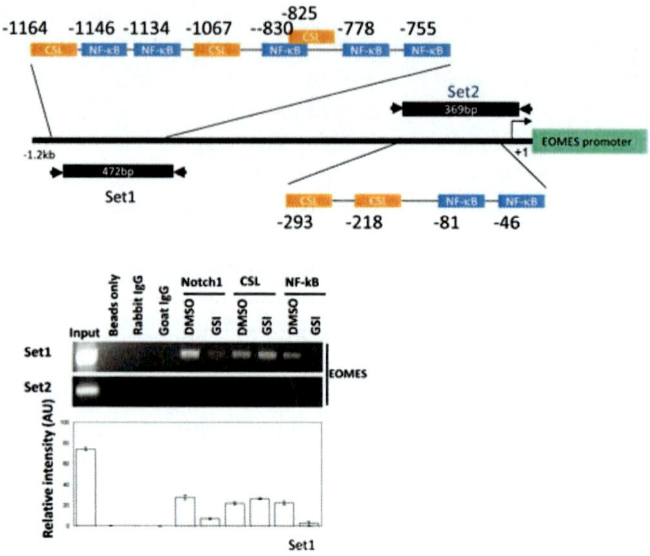

Fig. 3 ChIP analysis of the *EOMES* promoter reveals the presence of Notch/NF-κB complexes. **a** Diagrammatic view of the murine *EOMES* promoter indicating potential CSL or NF-κB sites and the design of two primer sets used for ChIP analysis. **b** Promoter sets 1 or 2 were used to amplify protein/DNA complexes isolated by ChIP. The data indicate that whereas Notch, CSL, and NF-κB proteins all bind to the *EOMES* promoter in the region amplified by primer set 1, only Notch and NF-κB binding is inhibited by GSI implying that CSL may bind constitutively to this region of the promoter while NF-κB binding requires cleavage of Notch

6 Biological Implications

The importance of recognizing that Notch signals in T cells both by canonical and non-canonical pathways in T cells is clear. As we describe above, Notch signaling plays an integral role in several key functions of mature peripheral T cells. In addition to its role in regulating peripheral T cell fate, dysregulated Notch expression may also play a critical role in disease. In this latter situation, it may be advantageous to block Notch signaling. Clearly, an understanding of how Notch signals in different cellular contexts is important to design appropriate tools to inhibit specific aspects of Notch signaling.

Many of the data describing a non-canonical, nonnuclear role for Notch in regulating peripheral T cell function are quite recent. Therefore, our understanding of the precise molecular mechanisms that mediate non-canonical, nonnuclear Notch function is in its infancy. New tools for the exploration of these novel non-canonical pathways will need to be developed if we are to fully understand the role Notch plays in regulating T cell function. Ultimately, an exploration of the various means by which Notch influences T cell function is likely to lead to the development of new methods to modulate Notch signaling. Such advances may allow us to manipulate this signaling pathway for the amelioration of disease.

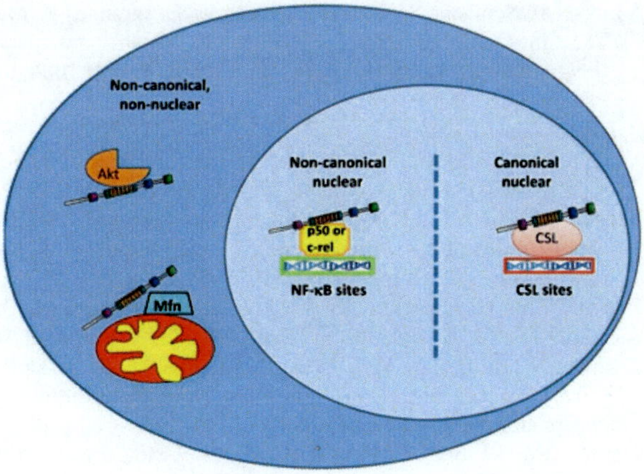

Fig. 4 Summary of possible canonical and non-canonical roles for Notch signaling in T cells

7 Conclusions

In conclusion, Notch signaling is a complicated and multifaceted process that can occur in the nucleus, in the cytosol, or in both compartments (summarized in Fig 4). As evidenced by the literature, Notch signaling in the nucleus can involve canonical CSL-dependent transcription. However, it is also clear that non-canonical nuclear signaling involving Notch and NF-κB family members exists, demonstrating another mechanism by which Notch regulates transcription. Additionally, data from several labs demonstrating nonnuclear functions for Notch open the possibility that, in $CD4^+$ T cells, Notch participates in cytosolic signaling pathways and these interactions may well influence cell fate and function. Future experimentation likely will fully resolve the various pathways used by Notch to mediate $CD4^+$ T cell fate decisions.

Acknowledgments We thank the members of the Minter and Osborne labs for their contributions and apologize to colleagues whose work was not cited in the interest of space. We particularly thank Richard Goldsby for his insightful comments. This work was supported by NIH AG025531.

References

Adler SH, Chiffoleau E, Xu L, Dalton NM, Burg JM, Wells AD, Wolfe MS, Turka LA, Pear WS (2003) Notch signaling augments T cell responsiveness by enhancing CD25 expression. J Immunol 171:2896–2903

Amsen D, Antov A, Jankovic D, Sher A, Radtke F, Souabni A, Busslinger M, McCright B, Gridley T, Flavell RA (2007) Direct regulation of Gata3 expression determines the T helper differentiation potential of Notch. Immunity 27:89–99

Amsen D, Blander JM, Lee GR, Tanigaki K, Honjo T, Flavell RA (2004) Instruction of distinct CD4 T helper cell fates by different notch ligands on antigen-presenting cells. Cell 117:515–526

Androutsellis-Theotokis A, Leker RR, Soldner F, Hoeppner DJ, Ravin R, Poser SW, Rueger MA, Bae SK, Kittappa R, McKay RD (2006) Notch signalling regulates stem cell numbers in vitro and in vivo. Nature 442:823–826

Auderset F, Schuster S, Coutaz M, Koch U, Desgranges F, Merck E, Macdonald HR, Radtke F, Tacchini-Cottier F (2012) Redundant Notch1 and Notch2 signaling is necessary for IFNgamma secretion by T Helper 1 cells during Infection with Leishmania major. PLoS Pathog 8:e1002560

Bassil R, Zhu B, Lahoud Y, Riella LV, Yagita H, Elyaman W, Khoury SJ (2011) Notch ligand delta-like 4 blockade alleviates experimental autoimmune encephalomyelitis by promoting regulatory T cell development. J Immunol 187:3380–3389

Benson RA, Adamson K, Corsin-Jimenez M, Marley JV, Wahl KA, Lamb JR, Howie SE (2005) Notch1 co-localizes with CD4 on activated T cells and Notch signaling is required for IL-10 production. Eur J Immunol 35:859–369

Cho OH, Shin HM, Miele L, Golde TE, Fauq A, Minter LM, Osborne BA (2009) Notch regulates cytolytic effector function in CD8 + T cells. J Immunol 182:3380–3389

Elyaman W, Bradshaw EM, Wang Y, Oukka M, Kivisakk P, Chiba S, Yagita H, Khoury SJ (2007) Jagged1 and delta1 differentially regulate the outcome of experimental autoimmune encephalomyelitis. J Immunol 179:5990–5998

Elzinga BM, Twomey C, Powell JC, Harte F, McCarthy JV (2009) Interleukin-1 receptor type 1 is a substrate for gamma-secretase-dependent regulated intramembrane proteolysis. J Biol Chem 284:1394–1409

Fang TC, Yashiro-Ohtani Y, Del Bianco C, Knoblock DM, Blacklow SC, Pear WS (2007) Notch directly regulates Gata3 expression during T helper 2 cell differentiation. Immunity 27:100–110

Fiorini E, Merck E, Wilson A, Ferrero I, Jiang W, Koch U, Auderset F, Laurenti E, Tacchini-Cottier F, Pierres M, Radtke F, Luther SA, Macdonald HR (2009) Dynamic regulation of notch 1 and notch 2 surface expression during T cell development and activation revealed by novel monoclonal antibodies. J Immunol 183:7212–7222

Giniger E (1998) A role for Abl in Notch signaling. Neuron 20:667–681

Haapasalo A, Kovacs DM (2011) The many substrates of presenilin/gamma-secretase. J Alzheimers Dis 25:3–28

Hamada Y, Kadokawa Y, Okabe M, Ikawa M, Coleman JR, Tsujimoto Y (1999) Mutation in ankyrin repeats of the mouse Notch2 gene induces early embryonic lethality. Development 126:3415–3424

Hansson ML, Popko-Scibor AE, Saint Just Ribeiro M, Dancy BM, Lindberg MJ, Cole PA, Wallberg AE (2009) The transcriptional coactivator MAML1 regulates p300 autoacetylation and HAT activity. Nucleic Acids Res 37:2996–3006

Hori K, Sen A, Kirchhausen T, Artavanis-Tsakonas S (2011) Synergy between the ESCRT-III complex and Deltex defines a ligand-independent Notch signal. J Cell Biol 195:1005–1015

Hsieh JJ-D, Henkel T, Salmon P, Robey E, Peterson MG, Hayward SD (1996) Truncated mammalian Notch1 activates CBF1/RBPJk-repressed genes by a mechanism resembling that of Epstein-Barr virus EBNA2. Mol Cell Biol 16:952–959

Jehn BM, Bielke W, Pear WS, Osborne BA (1999) Cutting edge: protective effects of notch-1 on TCR-induced apoptosis. J Immunol 162:635–638

Jin B, Shen H, Lin S, Li JL, Chen Z, Griffin JD, Wu L (2010) The mastermind-like 1 (MAML1) co-activator regulates constitutive NF-kappaB signaling and cell survival. J Biol Chem 285:14356–14365

Keerthivasan S, Suleiman R, Lawlor R, Roderick J, Bates T, Minter L, Anguita J, Juncadella I, Nickoloff BJ, Le Poole IC, Miele L, Osborne BA (2011) Notch signaling regulates mouse and human Th17 differentiation. J Immunol 187:692–701

Klueg KM, Muskavitch Marc MAT (1999) Ligand-receptor interactions and trans-endocytosis of delta, serrate and Notch: members of the Notch signalling pathway in Drosophila. J Cell Sci 112:3289–3297

Le Gall M, De Mattei C, Giniger E (2008) Molecular separation of two signaling pathways for the receptor, Notch. Dev Biol 313:556–567

Lee SH, Wang X, DeJong J (2000) Functional interactions between an atypical NF-kappaB site from the rat CYP2B1 promoter and the transcriptional repressor RBP-Jkappa/CBF1. Nucleic Acids Res 28:2091–2098

Liu Z, Teng L, Bailey SK, Frost AR, Bland KI, LoBuglio AF, Ruppert JM, Lobo-Ruppert SM (2009) Epithelial transformation by KLF4 requires Notch1 but not canonical Notch1 signaling. Cancer Biol Ther 8:1840–1851

Mann J, Oakley F, Johnson PW, Mann DA (2002) CD40 induces interleukin-6 gene transcription in dendritic cells: regulation by TRAF2, AP-1, NF-kappa B, AND CBF1. J Biol Chem 277:17125–17138

McElhinny AS, Li JL, Wu L (2008) Mastermind-like transcriptional co-activators: emerging roles in regulating cross talk among multiple signaling pathways. Oncogene 27:5138–5147

Minter LM, Turley DM, Das P, Shin HM, Joshi I, Lawlor RG, Cho OH, Palaga T, Gottipati S, Telfer JC, Kostura L, Fauq AH, Simpson K, Such KA, Miele L, Golde TE, Miller SD, Osborne BA (2005) Inhibitors of gamma-secretase block in vivo and in vitro T helper type 1 polarization by preventing Notch upregulation of Tbx21. Nat Immunol 6:680–688

Mukherjee T, Kim WS, Mandal L, Banerjee U (2011) Interaction between Notch and Hif-alpha in development and survival of Drosophila blood cells. Science 332:1210–1213

Palaga T, Miele L, Golde TE, Osborne B (2003) TCR-mediated Notch signaling regulates proliferation and IFN-gamma production in peripheral T cells. J Immunol 171:3019–3024

Palaga T and Minter LM (2012) Notch signaling and its emerging role in autoimmunity. Front Biol. doi:10.1007/s11515-012-1209-z (epub ahead of print)

Perumalsamy LR, Nagala M, Banerjee P, Sarin A (2009) A hierarchical cascade activated by non-canonical Notch signaling and the mTOR-Rictor complex regulates neglect-induced death in mammalian cells. Cell Death Differ 16:879–889

Perumalsamy LR, Nagala M, Sarin A (2010) Notch-activated signaling cascade interacts with mitochondrial remodeling proteins to regulate cell survival. Proc Natl Acad Sci U S A 107:6882–6887

Raafat A, Lawson S, Bargo S, Klauzinska M, Strizzi L, Goldhar AS, Buono K, Salomon D, Vonderhaar BK, Callahan R (2009) Rbpj conditional knockout reveals distinct functions of Notch4/Int3 in mammary gland development and tumorigenesis. Oncogene 28:219–230

Reynolds ND, Lukacs NW, Long N, Karpus WJ (2011) Delta-like 4 regulates central nervous system T cell accumulation during experimental autoimmune encephalomyelitis. J Immunol 187:2803–2813

Rutz S, Mordmüller B, Sakano S, Scheffold A (2005) Notch ligands Delta-like1, Delta-like4 and Jagged1 differentially regulate activation of peripheral T helper cells. Eur J Immunol 35:2443–2451

Sade H, Krishna S, Sarin A (2004) The anti-apoptotic effect of Notch-1 requires p56*lck*-dependent, Akt/PKB-mediated signaling in T cells. JCB 279:2937–2944

Saxena MT, Schroeter EH, Mumm JS, Kopan R (2001) Murine Notch homologs (N1–4) undergo presenilin-dependent proteolysis. JBC 276:40268–40273

Seo W, Ziltener HJ (2009) CD43 processing and nuclear translocation of CD43 cytoplasmic tail are required for cell homeostasis. Blood 114:3567–3577

Shawber C, Nofziger D, Hsieh JJ, Lindsell C, Bogler O, Hayward D, Weinmaster G (1996) Notch signaling inhibits muscle cell differentiation through a CBF1-independent pathway. Development 122:3765–3773

Shen H, McElhinny AS, Cao Y, Gao P, Liu J, Bronson R, Griffin JD, Wu L (2006) The Notch coactivator, MAML1, functions as a novel coactivator for MEF2C-mediated transcription and is required for normal myogenesis. Genes Dev 20:675–688

Shin HM, Minter LM, Cho OH, Gottipati S, Fauq AH, Golde TE, Sonenshein GE, Osborne BA (2006) Notch1 augments NF-kappaB activity by facilitating its nuclear retention. EMBO J 25:129–138

Swiatek PJ, Lindsell CE, del Amo FF, Weinmaster G, Gridley T (1994) Notch1 is essential for postimplantation development in mice. Genes Dev 8:707–719

Takeichi N, Yanagisawa S, Kaneyama T, Yagita H, Jin YH, Kim BS, Koh CS (2010) Ameliorating effects of anti-Dll4 mAb on Theiler's murine encephalomyelitis virus-induced demyelinating disease. Int Immunol 22:729–738

Tanigaki K, Tsuji M, Yamamoto N, Han H, Tsukada J, Inoue H, Kubo M, Honjo T (2004) Regulation of alphabeta/gammadelta T cell lineage commitment and peripheral T cell responses by Notch/RBP-J signaling. Immunity 20:611–622

Tao X, Constant S, Jorritsma P, Bottomly K (1997) Strength of TCR signal determines the costimulatory requirements for Th1 and Th2 CD4 + T cell differentiation. J Immunol 159:5956–5963

Vales LD, Friedl EM (2002) Binding of C/EBP and RBP (CBF1) to overlapping sites regulates interleukin-6 gene expression. J Biol Chem 277:42438–42446

van Tetering G, van Diest P, Verlaan I, van der Wall E, Kopan R, Vooijs M (2009) Metalloprotease ADAM10 is required for Notch1 site 2 cleavage. JBC 284:31018–31027

Vied C, Kalderon D (2009) Hedgehog-stimulated stem cells depend on non-canonical activity of the Notch co-activator mastermind. Development 136:2177–2186

Wongchana W, Palaga T (2012) Direct regulation of interleukin-6 expression by Notch signaling in macrophages. Cell Mol Immunol 9:155–162

Wu L, Aster JC, Blacklow SC, Lake R, Atravanis-Tsakonas S, Griffin JD (2000) MAML1, a human homologue of Drosophila mastermind, is a transcriptional co-activator for NOTCH receptors. Nat Genet 26:484–489

Zecchini V, Brennan K, Martinez-Arias A (1999) An activity of Notch regulates JNK signalling and affects dorsal closure in Drosophila. Curr Biol 9:460–469

Zhao Y, Katzman RB, Delmolino LM, Bhat I, Zhang Y, Gurumurthy CB, Germaniuk-Kurowska A, Reddi HV, Solomon A, Zeng MS, Kung A, Ma H, Gao Q, Dimri G, Stanculescu A, Miele L, Wu L, Griffin JD, Wazer DE, Band H, Band V (2007) The notch regulator MAML1 interacts with p53 and functions as a coactivator. J Biol Chem 282:11969–11981

The Role of Notch in the Differentiation of CD4$^+$ T Helper Cells

Floriane Auderset, Manuel Coutaz and Fabienne Tacchini-Cottier

Abstract CD4$^+$ T helper cells are playing critical roles in host defense to pathogens and in the maintenance of immune homeostasis. Na CD4$^+$ T cells, upon antigen-specific recognition, receive signals to differentiate into distinct effector T helper cell subsets characterized by their pattern of cytokine production and specific immune functions. A tight balance between these different subsets ensures proper control of the immune response. There is increasing evidence revealing an important role for Notch signaling in the regulation of CD4$^+$ T helper cell differentiation or function in the periphery. However, the exact mechanisms involved remain unclear and appear contradictory. In this review, we summarize current knowledge and discuss recent advances in the field to reconcile different views on the role of Notch signaling in the differentiation of functional T helper subsets.

Abbreviations

Ab	Antibody
Ag	Antigen
APC	Antigen-presenting cells
DC	Dendritic cell
DNMAML	Dominant negative mastermind/Lag3
Dll	Delta-like ligand
EAE	Experimental autoimmune encephalomyelitis
GSI	γ-Secretase inhibitor
IFN	Interferon
IL-...	Interleukin-...

F. Auderset · M. Coutaz · F. Tacchini-Cottier (✉)
Department of Biochemistry, WHO Immunology Research and Training Center,
University of Lausanne, Chemin Des Boveresses 155, 1066 Epalinges, Switzerland
e-mail: Fabinne.Tacchini-Cottier@unil.ch

Current Topics in Microbiology and Immunology (2012) 360: 115–134
DOI: 10.1007/82_2012_227

Published Online: 1 June 2012

iTreg	Inducible T regulatory
LPS	Lipopolysaccharid
MAML	Mastermind/Lag3
NF-κB	Nuclear factor kappaB
NICD	Notch intracellular domain
PGE2	Prostaglandin-E2
RBP-Jκ	Recombination signal-binding protein-Jκ
STAT	Signal transducer and activator of transcription
Th	T helper
Treg	T regulatory

Contents

1 Introduction

In response to a vast array of microbial pathogens, innate activated cells induce na CD4$^+$ T cells to differentiate into distinct T helper (Th) cells. Depending, among other factors, on the signals they receive from the environment, each subset will provide the specific help needed for other cells to eradicate the pathogens encountered. These distinct CD4$^+$ Th subsets with specific cytokine profiles orchestrate the adaptive immune response toward the many microbial pathogens encountered. Failure to control proper CD4$^+$ Th cell function may lead to diseases

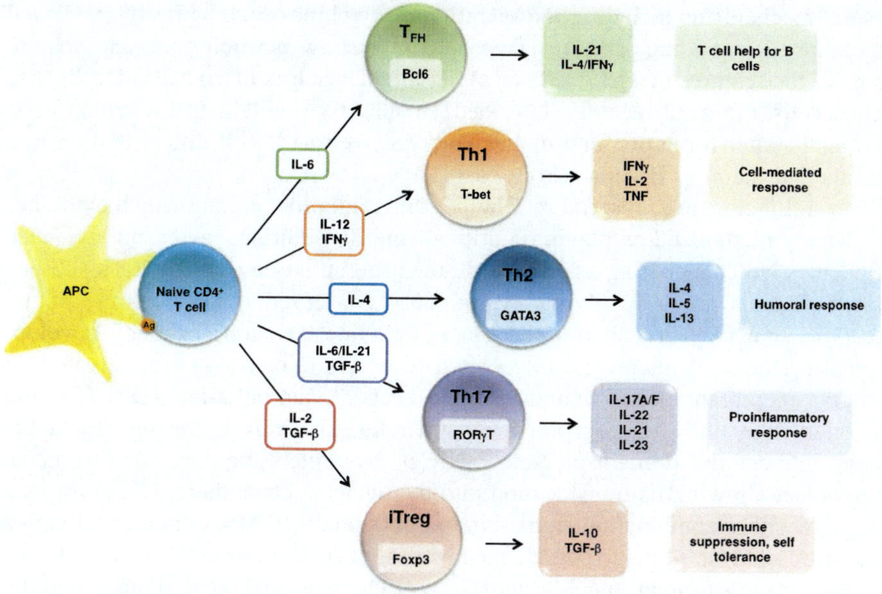

Fig. 1 Classical schematic representation of CD4$^+$ helper T cell differentiation. Upon activation by their cognate antigen presented by professional antigen presenting cells, CD4$^+$ T cells differentiate into various specialized T helper cell subsets, characterized by distinct patterns of cytokine secretion. The different sets of cytokines they secrete give each of these lineages a specific function, allowing the development of an appropriate immune response against pathogens of various types. The polarization of CD4$^+$ T cells is regulated by several factors including the cytokine environment, the type of antigen, the strength, and nature of costimulatory signals given by APCs and the affinity of peptide/MHC II complex with the TCR

or excessive inflammation. The understanding of the mechanisms regulating the differentiation and function of these Th subsets is therefore of utmost interest.

The specific characteristics of the main Th subsets are summarized in Fig. 1. Th1 cells facilitate the clearance of intracellular pathogens by secreting IFNγ, their signature cytokine. Th2 cells secrete the cytokines IL-4, IL-5, and IL-13 and their development is linked with the clearance of extracellular parasites and allergies. Th17 cells contribute to the mucosal host defense against certain bacterial and fungal infections, and are also involved in autoimmune and inflammatory diseases. IL-17 is their signature cytokine. Although each specific subset was initially thought to be a terminally differentiated fixed lineage, some plasticity can exist between these subsets (Locksley 2009; O'Shea and Paul 2010). For instance, Th17 cells show more plasticity than Th1 and Th2 cells, being able to switch off their cytokine signature pattern and acquire new ones such as that of regulatory T (Treg) cells (reviewed in Hirota et al. 2011). Regulatory T cells that arise from peripheral CD4$^+$ T cells (iTreg) are found in high numbers in various lymphoid or non-lymphoid tissues, including inflamed tissues and the intestinal mucosa. They restrain excessive response by other Th subsets preventing tissue damage through several mechanisms including the secretion of cytokines such as IL-10 (reviewed in Campbell and Koch (2011)). iTreg cells also show plasticity and can acquire some of the features of Th17 cells or of T follicular helper (Tfh) cells. T follicular helper cells, the most recently described Th subset, provide help for lymph node germinal center formation and maintenance, as well as B cell differentiation into plasma and memory B cells (Crotty 2011).

Several factors are involved in T helper differentiation, among which cytokines are widely recognized as playing a critical role. In addition, increasing evidence implicates Notch signaling as a player in the differentiation of distinct Th subsets. Notch proteins are a family of evolutionary conserved transmembrane bound receptors that play crucial roles in binary cell fate decisions in many developmental systems. In mammals, four Notch receptors (Notch1–4) are activated by five transmembrane bound ligands (Jagged 1, and 2, and delta-like (Dll) 1, 3, and 4). Interaction of Notch receptors with their ligands initiates the release of the active intracellular domain of Notch (NICD) by proteolytic cleavages from the membrane, allowing its translocation into the nucleus. Once there, NICD forms a complex with recombination signal-binding protein-Jκ (RBP-Jκ) and coactivators such as mastermind-like-1 (MAML), converting it from a repressor to an activator of transcription (Sandy and Maillard 2009). The role of Notch signaling in Th differentiation is not clear (Amsen et al. 2009; Radtke et al. 2010). Some studies show that Notch supports the development of Th1 cells, whereas others report a role for Notch in Th2, Th17, or iTreg cell differentiation. Furthermore, the recently described plasticity of some of these Th subsets complicates some interpretations. This review summarizes and discusses findings on the role of Notch signaling in Th cell responses, and highlights certain areas of ambiguity.

2 The Role of Notch Signaling in CD4$^+$ (Th1 Cell Differentiation)

2.1 Delta-Like Ligands Preferentially Induce the Activation of CD4$^+$ Th1 Cells

The first indication of a contribution of Notch in Th1 differentiation was obtained in vitro with experiments showing that expression of Delta-like ligands but not that of Jagged ligands by antigen presenting cells (APCs) correlated with their ability to induce a Th1 response. In vitro exposure of APCs to various Th1-inducing stimuli such as the TLR agonists LPS (Skokos and Nussenzweig 2007; Amsen et al. 2007; Napolitani et al. 2005), CpG, *Propionibacterium acnes* (Sun et al. 2008; Worsley et al. 2008), Poly(I:C) (Napolitani et al. 2005) or infection with the respiratory syncytial virus (RSV) (Schaller et al. 2007) or the influenza H1N1 virus (Ito et al. 2011) resulted in Dll upregulation on dendritic cells (DCs) or macrophages which correlated with the induction of Th1 cell differentiation. Several

TLR agonists could further synergize in driving Th1 differentiation by increasing Dll4 expression on APCs (Napolitani et al. 2005). Conversely, deficiency in the TLR adaptor protein MyD88 which prevented the induction of Dll expression on APCs inhibited the DC-mediated induction of Th1 cells (Amsen et al. 2004; Rudd et al. 2007; Sun et al. 2008). In addition, CD4^{+} T cells cocultured with APCs that ectopically express Dll1 or Dll4 acquired a Th1 profile, secreting high levels of IFNγ (Amsen et al. 2004; Sun et al. 2008; Skokos and Nussenzweig 2007). To separate the direct effect of Dll on Th1 differentiation from that coming from other signals provided by activated dendritic cells, CD4^{+} T cells were directly exposed to Dll1-Fc fusion protein. Under in vitro conditions promoting Th1 cell differentiation, the presence of Dll1-Fc fusion protein enhanced IFNγ secretion by CD4^{+} T cells, whereas it impaired secretion of IL-4 under Th2-inducing conditions. Interestingly, Dll appeared to use an RBP-Jκ-independent pathway to promote Th1 cell differentiation (Amsen et al. 2004).

The ability of Dll proteins to promote a Th1 cell response was further demonstrated in vivo. Administration of a Notch-activating Dll1-Fc fusion protein in *Leishmania major*-infected BALB/c mice promoted the development of IFNγ-secreting Th1 cells, conferring them ability to control the infection, despite their susceptible genetic background (Maekawa et al. 2003). Conversely, treatment of *L. major*-infected C57BL/6 mice with a Dll1-mutated Fc fusion protein, which blocks Notch signaling by competing with endogenous Dll1, resulted in enhanced susceptibility despite their resistant genetic background. The susceptibility of these mice was correlated with reduced IFNγ secretion (Maekawa et al. 2003). Similarly, neutralization of Dll1 during infection with influenza H1N1 virus induced higher mortality, decreased levels of IFNγ and impaired viral clearance (Ito et al. 2011). In the same line, in vivo inhibition of Dll4-mediated Notch activation in IL-12p40$^{-/-}$ mice resulted in impaired development of OVA-specific Th1 cells in response to LPS and OVA, revealing that Dll4 induces the differentiation of Th1 cells in an IL-12-independent way in this experimental setting (Skokos and Nussenzweig 2007).

The mechanisms promoting Th1 differentiation by Dll triggering remain to be defined. Induction of Dll1 and/or Dll4 expression impairs the development of a Th2 response, a process thought to allow the development of a Th1 response “by default” (Sun et al. 2008; Okamoto et al. 2008; Schaller et al. 2007). This was first demonstrated in vitro. CD4^{+} T cells cultured in presence of exogenous IL-4 and activated DCs showed decreased IL-4 secretion in presence of recombinant Dll4, a process independent of IFNγ or T-bet (Sun et al. 2008). The ability of Dll1 to antagonize the Th2 response was further reported in vivo in the model of asthma, a Th2-mediated allergic disease. Administration of Dll1-Fc fusion protein in sensitized mice prevented airway responsiveness, which was correlated with decreased Th2 effector cytokine secretion and increased IFNγ levels (Okamoto et al. 2008). Another study showed that upon RSV infection, the development of a protective Th1 response was biased toward a Th2 type of response in mice treated with an anti-Dll4 mAb, resulting in increased airway hyper-reactivity, decreased IFNγ secretion and increased IL-13 and IL-5 cytokine secretion by Th2 cells (Schaller et al. 2007).

Collectively, the in vitro and in vivo studies discussed here show that signaling resulting from the engagement of Dll has a major impact on the development of functional Th1 cells. This process appears to be indirect, repressing the development of Th2 (or as discussed below, that of Th17 cells), thereby allowing Th1 development apparently by default.

2.2 A Preponderant Role for Notch1 Signaling in Driving the Differentiation of IFNγ-Secreting CD4⁺ Th1 Cells

Some controversies remain regarding the identity of the Notch receptors driving Dll-induced Th1 responses. Transfection of $CD4^+$ T cells with the Notch3 intracellular domain (N3ICD) resulted in enhanced IFNγ secretion accompanied by lower levels of IL-4, whereas transfection of N1ICD did not affect neither IFNγ nor IL-4 secretion (Maekawa et al. 2003). In contrast, in another study, the transfection of N1ICD led to T-bet mRNA expression and increased IFNγ secretion by $CD4^+$ T cells (Minter et al. 2005). The nature of the T cells transfected [lymph node $CD4^+$ T cells (Maekawa et al. 2003) vs. T cell hybridoma cell line (Minter et al. 2005)], might explain the discrepancies observed. Treatment of $CD4^+$ T cells with an inhibitor of γ-secretase (GSI), which blocks the Notch pathway independently of the receptor or ligand involved, reduced IL-12-mediated Th1 polarization in vitro (Minter et al. 2005). However, no effect on Th1 cell development was reported in other in vitro studies that used genetic loss-of-function approaches such as inactivation of Notch1 and/or Notch2 or of RBP-Jκ in T cells (Amsen et al. 2004, 2007; Tacchini-Cottier et al. 2004). Similarly, inactivation of both presenilin1 and 2, two major components of the γ-secretase complex, did not result in a measurable impact on the differentiation of $CD4^+$ Th1 cells in vitro (Ong et al. 2008).

Although experiments using $CD4^+$ T cells with genetic inactivation of Notch indicated that Notch is dispensable for the development of Th1 cells in vitro, Notch signaling appeared essential for the development of IFNγ-secreting Th1 cells in vivo. We recently showed that expression of either Notch1 or Notch2 is necessary for the differentiation of functional $CD4^+$ Th1 cells and the consequent resolution of *L. major* infection (Auderset et al. 2012). While mice deficient for both Notch1 and Notch2 were highly susceptible to infection with *L. major*, single deletion of Notch1 (Tacchini-Cottier et al. 2004) or Notch2 (Auderset et al. 2012) had no effect on the development of an efficient protective Th1 response to *L. major*, indicating a functional redundancy between Notch1 and Notch2 in this process. The inability of mice deficient for both Notch1 and Notch2 in their T cells to develop IFNγ-secreting Th1 cells suggests that Notch3 is not functionally redundant and cannot overcome the absence of Notch1 and Notch2 during *L. major* infection. In contrast, in different experimental settings, Notch3 was shown to be the major receptor driving Th1 responses both in vitro and in vivo (Maekawa et al. 2003; Jurynczyk et al. 2008). Notch3 has been reported to be a

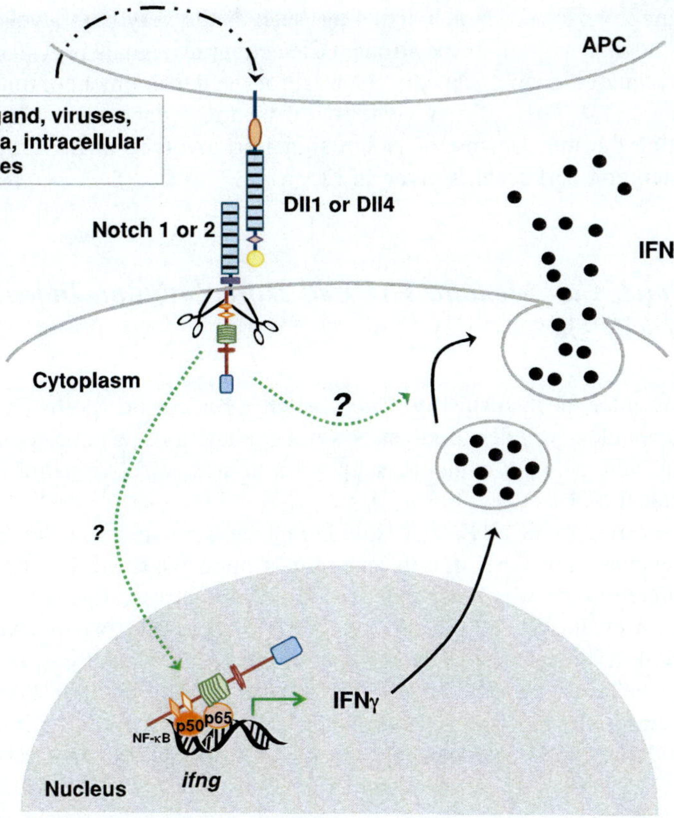

Fig. 2 The role of Notch in Th1 cell differentiation. Upon stimulation with Th1-inducing stimuli, APCs preferentially express Dll1 or Dll4. Dll proteins signal through Notch1 and in a lesser way through Notch2 to promote the development of functional IFNγ-secreting cells most likely independently of RBP-Jκ. The RBP-Jκ-independent secretion of IFNγ might occur through interaction between NICD and the nuclear factor of NF-κB pathway p50 and p65. Furthermore, Notch is essential for IFNγ secretion, but the mechanisms by which Notch can induce delivery of IFNγ remains to be identified

Notch1 target gene in thymocytes (van den Brandt et al. 2006), however this is controversial (Shi et al. 2011). Thus, one cannot exclude that Notch3, if expressed downstream of Notch1 signaling, could synergize with Notch1 in driving Th1 differentiation, without being essential for the initiation or maintenance of the response. However, Notch3 expression was not detectable on $CD4^+$ T cells of wild type mice or mice deficient in Notch1 and/or Notch2 in their T cells neither by mRNA nor at the protein level during *L. major* infection (Auderset et al. 2012). Thus, the role of Notch3 in driving the development of functional Th1 cells during infection remains to be established. Individual Notch usage may vary in response to different pathogens. Altogether, the involvement of Notch receptor triggering appears dispensable during in vitro differentiation, perhaps due to factors bypassing the need for Notch signaling, such as the very high levels of cytokines present in vitro, as well as the strong TCR activating signals provided. In contrast, Notch receptor signaling appears essential for the development of functional IFNγ-secreting $CD4^+$ Th1 cells in vivo. Based on these data, an hypothetical model suggesting the mechanisms of Notch signaling involved in the differentiation of IFNγ-secreting Th1 cells is given in Fig. 2.

2.3 Notch Can Mediate Th1 Cell Differentiation Independently of RBP-Jk

The molecular mechanisms by which Notch signaling drives the Th1 differentiation are not clearly understood yet. Several explanations have been proposed in the literature and growing evidence supports a noncanonical signaling pathway (i.e. independent of RBP-Jκ).

Overexpression of N1ICD or N3ICD in T cells resulted in activation of *Tbx21*, which encodes for T-bet, the master transcription factor of Th1 cells. However, Th1 differentiation was independent of IL-12, the classical activator of Th1 cells (Maekawa et al. 2003; Minter et al. 2005; Skokos and Nussenzweig 2007). In support for a direct role of Notch signaling on *Tbx21* regulation, several RBP-Jκ-binding sites were identified on the *Tbx21* gene promoter, within 1 kb upstream of the transcription start codon. Both Notch1 and RBP-Jκ could be immunoprecipitated with the *Tbx21* promoter region isolated from hybridoma T cells, suggesting that Notch can directly induce T-bet expression (Minter et al. 2005). However, the status of T-bet as a direct target of Notch signaling has been challenged in other studies: binding of RBP-Jκ or Notch on the *Tbx21* promoter region was not detectable in primary cells (Fang et al. 2007). In addition, two other independent studies performed in vivo argue against the involvement of RBP-Jκ in the regulation of T-bet. Mice which conditionally express a dominant negative MAML protein, thereby deprived of RBP-Jκ-mediated transcriptional activation, as well as mice specifically deficient for RBP-Jκ in their T cells were able to develop an efficient Th1 immune response and thereby resolve their infection against *L. major* (Tu et al. 2005; Auderset et al. 2012). In addition, an upregulation of T-bet and IFNγ mRNA expression in $CD4^+$ T cells deficient for both Notch1 and Notch2 was reported, arguing against a direct activation of T-bet by Notch signaling (Auderset et al. 2012). Defective secretion but not transcription of IFNγ was measured in Notch-deficient $CD4^+$ T cells, suggesting that Notch signaling regulates the secretion of IFNγ by Th1 cells (Auderset et al. 2012).

The mechanisms involved in RBP-Jκ-independent Notch signaling are complex and not yet clearly defined (Sanalkumar et al. 2010). Numerous studies have reported a regulation of NF-κB by Notch through different mechanisms including RBP-Jκ-independent (noncanonical) pathway (Osipo et al. 2008). In line with these data, Notch has been found to associate with the *ifng* promoter despite the absence of a consensus binding sequence for RBP-Jκ (Shin et al. 2006). This interaction is thought to occur through association of NICD with the NF-κB factors p50 and p65, resulting in both nuclear retention of p50 and p65 and induction of IFNγ (Shin et al. 2006).

Collectively, in vivo and in vitro data support the concept of a RBP-Jκ-independent function of Notch in the development of functional IFNγ-secreting Th1 cells.

3 The Role of Notch Signaling in the Differentiation of CD4$^+$ Th2 Cells

3.1 Jagged Ligands Induce Preferentially the Development of CD4$^+$ Th2 Cells

In contrast to the role of Dll in Th1 differentiation, APCs that express Jagged ligands preferentially induced the development of Th2 cells. Stimuli such as *Schistosoma* egg antigens (SEA), cholera toxin, prostanglandin E2 (Amsen et al. 2004; Krawczyk et al. 2008; Worsley et al. 2008), or allergens such as dust mite extracts (Krishnamoorthy et al. 2008) were shown to induce Jagged2 expression by APCs. Jagged1 was found to be upregulated by DCs following OVA stimulation (Okamoto et al. 2009) or LPS stimulation similarly to Dll4 (Amsen et al. 2004). LPS is a strong Th1-inducer but can also promote a Th2 response in absence of MyD88. Interestingly, MyD88$^{-/-}$ DCs exposed to LPS showed upregulation of Jagged1 but not Dll4, indicating that Jagged1 expression correlates with the ability of LPS to promote Th2 differentiation while Dll4 expression correlates with the LPS-induced Th1 differentiation (Amsen et al. 2004). Immature human myeloid DCs constitutively express Jagged1, which confers them the ability to induce a CD4$^+$ Th2 response, as revealed by the induction of GATA3 expression and IL-4 secretion by CD4$^+$ T cells. In addition, upon TLR ligation of these cells, Jagged1 expression was reduced, concomitantly with the upregulation of Dll4 and a switch from Th2 to Th1 differentiation (Liotta et al. 2008). In further support of a role for Jagged in promoting Th2 cell differentiation, CD4$^+$ T cells cultured in vitro in presence of DCs that ectopically express Jagged1 acquired a Th2 phenotype characterized by high levels of IL-4 and IL-5 secretion (Amsen et al. 2004). Conversely, downregulation of Jagged1 by mRNA interference impaired the ability of human myeloid DCs to induce a Th2 differentiation (Liotta et al. 2008) and Jagged2 deficiency reduced the Th2 differentiation promoted by SEA-primed DCs in vitro (Amsen et al. 2004). Moreover, GSI pretreatment of CD4$^+$ T cells cocultured with OVA-pulsed DCs expressing Jagged1 resulted in a markedly reduced Th2 cytokine secretion by these CD4$^+$ T cells (Okamoto et al. 2009). In sharp contrast, in two other studies, neither suppression nor forced expression of Jagged2 by DCs affected their ability to mediate Th2 cell differentiation in vitro and in vivo (Worsley et al. 2008; Krawczyk et al. 2008). Given that, both Jagged1 and Jagged2 expression favor the development of Th2 cells in vitro redundant functions between both ligands cannot be excluded and might explain the lack of effect of Jagged2 suppression observed in these latter studies. Alternatively, specific antigens may induce selectively one of the two Jagged ligands.

Data obtained in vivo support a contribution of Jagged ligands in driving Th2 differentiation. Administration of stimulatory Jagged1-Fc fusion protein induced a Th2 type immune response in the mouse model of experimental autoimmune encephalomyelitis (EAE), resulting in delayed disease onset and attenuation of the symptoms (Elyaman et al. 2007). Similarly, treatment of allergic asthma-sensitized mice with Notch-stimulatory Jagged1-Fc fusion protein enhanced airway hyperresponsiveness (AHR), a process resulting from the accumulation of Th2 cytokine-secreting $CD4^+$ T cells in the lungs (Okamoto et al. 2009). Conversely, inhibition of Jagged1 expression with small interfering RNA or treatment with GSI prevented AHR and airway inflammation (Okamoto et al. 2009).

Thus, the majority of the data support a role for Jagged ligands in favoring the differentiation of Th2 cells.

3.2 Role of Notch in Th2 Cell Differentiation

Notch signaling is not required for the differentiation of $CD4^+$ Th2 cells in vitro, as shown in experiments using $CD4^+$ T cells deficient for Notch1 and/or Notch2 (Amsen et al. 2007; Tacchini-Cottier et al. 2004; unpublished data), for presenilin1 and presenilin2 (Ong et al. 2008), or expressing DNMAML (Tu et al. 2005). Pretreatment of na $CD4^+$ T cells with GSI also prevented Th2 differentiation in vitro (Minter et al. 2005).

In sharp contrast, Notch signaling appears essential in the differentiation of IL-4-secreting Th2 cells in vivo. Pharmacological inhibition of Notch with GSI in the mouse model of allergic asthma reduced the disease symptoms, in correlation with a decreased Th2 cytokine secretion (Okamoto et al. 2009; Kang et al. 2009). These results were further confirmed in mice with conditional inactivation of Notch signaling in $CD4^+$ T cells. Mice that conditionally express a dominant negative MAML in their T cells that prevents Notch signaling were unable to mount a protective Th2 type of immune response in response to infection with *Trichuris muris* (Tu et al. 2005). In the same vein, Th2 differentiation was abrogated in mice deficient for RBP-Jκ in their T cells upon immunization with SEA (Amsen et al. 2007) and with antigens (OVA) adsorbed to alum (Tanaka et al. 2006). Moreover, a Th2 response was also defective in mice deficient for both Notch1 and Notch2 expression on their T cells (Amsen et al. 2007). Of note, most of these studies correlated a defective Th2 response with impaired IL-4 secretion. We recently showed that in absence of Notch1 and Notch2 expression, $CD4^+$ T cells did not secrete Il-4 but secreted high levels of the Th2 cytokines IL-13 and IL-5 (Auderset et al. 2012).

Altogether, these data indicate that Notch signaling is essential for the development of IL-4-secreting CD4$^+$ T cells in vivo, but may be dispensable for that of IL-13- or IL-5-secreting Th2 cells.

3.3 IL-4 and GATA3 Promoters are Direct Targets of Notch Signaling

The finding that RBP-Jκ-mediated Notch signaling is essential for Th2 differentiation and IL-4 secretion in vivo (Amsen et al. 2007; Tu et al. 2005; Tanaka et al. 2006) suggested that Notch might act by direct activation of Th2-related genes. Several RBP-Jκ-binding sites were identified in the 3′ enhancer region of the *il4* locus (known as CNS-2 or HS5) (Amsen et al. 2004; Tanaka et al. 2006). Supporting the idea of a direct control of Notch on IL-4 expression, retroviral transfection of N1ICD enhanced IL-4 secretion by CD4$^+$ T cells (Fang et al. 2007). Chromatin immunoprecipitation experiments showed that both Notch1 and RBP-Jκ were able to bind CNS-2 in CD4$^+$ T cells in vitro (Fang et al. 2007) and two other studies further confirmed the responsiveness of CNS-2 to Notch in vivo (Amsen et al. 2004; Tanaka et al. 2006). In line with these data, the expression of a reporter gene under the control of CNS-2 was abrogated in RBP-Jκ-deficient CD4$^+$ T cells (Tanaka et al. 2006). Collectively, these data demonstrate that Notch regulates IL-4 expression through interaction with CNS-2-binding RBP-Jκ. The finding that IL-4 regulation occurs downstream of Notch signaling likely explains why addition of exogenous IL-4 can override the need for Notch signaling during Th2 differentiation in vitro (Tacchini-Cottier et al. 2004; Amsen et al. 2007; Minter et al. 2005; Ong et al. 2008).

IL-4 signals through the STAT6 transcription factor, which in turn induces the expression of GATA3, the master transcriptional regulator of Th2 differentiation (Zheng and Flavell 1997). Two studies showed that Notch could induce Th2 cell development in mice deficient for STAT6 through a mechanism dependent on GATA3 (Amsen et al. 2007; Fang et al. 2007). These observations suggested that in addition to IL-4, GATA3 might also be directly regulated by Notch, independently of STAT6. Supporting this hypothesis, GATA3 expression was strongly upregulated in STAT6$^{-/-}$ CD4$^+$ T cells transfected with N1ICD (Amsen et al. 2007; Fang et al. 2007). *Gata3* gene expression is controlled by two promoters, localized around 10 kb apart (Asnagli et al. 2002). Each of these promoters drives the expression of exon 1a or 1b, followed by a second common exon. A RBP-Jκ-binding site was found on the most upstream promoter of GATA3, consistent with the strong induction of exon 1a, but not exon 1b, upon Notch signaling (Amsen et al. 2007; Fang et al. 2007). Chromatin immunoprecipitation further confirmed endogenous binding of Notch1 and RBP-Jκ to this site, supporting a regulatory role for Notch signaling in GATA3 expression (Amsen et al. 2007; Fang et al. 2007).

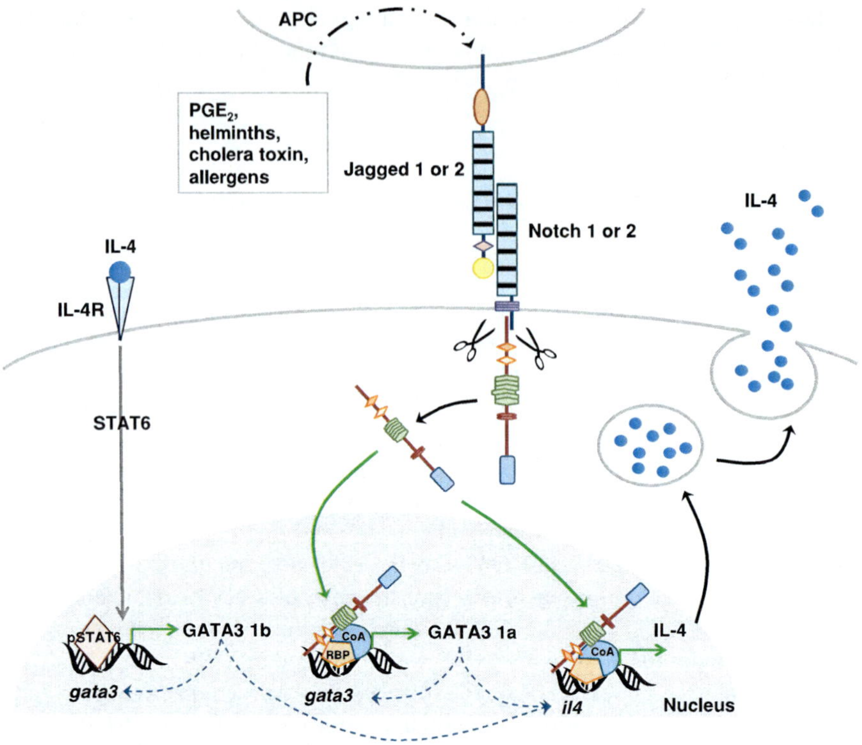

Fig. 3 The role of Notch in Th2 cell differentiation. Jagged ligands are preferentially expressed by APCs under Th2-inducing conditions. Upon activation by Jagged ligands, Notch1 or Notch2 signal via RBP-Jκ to promote GATA3 exon 1a and IL-4 expression. In turn, GATA3 enhances its own expression and induces IL-4 expression, stabilizing the Th2 phenotype. The expression of GATA3 exon 1b downstream of IL-4 signaling might bypass the need of Notch in Th2 differentiation

Altogether, these data provide two mechanisms by which Notch can regulate the development of Th2 cells. Notch signaling can directly induce IL-4 expression, which further drives Th2 polarization, a mechanism which may be important when IL-4 is absent or available only at low levels. It is interesting to note the close proximity between GATA3- and RBP-Jκ-binding sites on the *il4* locus and the synergistic activity of GATA3 and Notch for IL-4 expression (Fang et al. 2007). It is conceivable that GATA3 induces chromatin remodeling in the *il4* locus, rendering the CNS-2 region more accessible to Notch.

Collectively, numerous data demonstrate the essential role of Notch in GATA3 transcription. However, the direct regulation of GATA3 exon 1a by Notch signaling may be context dependent, and when expression of GATA-3 is driven by exon 1b, Notch may be dispensable for Th2 differentiation. Consistent with this hypothesis, we have recently shown that in absence of Notch1 and Notch2, CD4$^+$ T cells failed to secrete IL-4 but secreted the Th2 cytokines IL-5 and IL-13 (Auderset et al. 2012). The levels of GATA3 1a and 1b mRNA expression in CD4$^+$ T cells were similar to that of control T cells (unpublished data), suggesting a context-dependent effect of Notch on GATA3 1a transcription. The mechanisms leading to alternative selection of GATA3 exon 1a or 1b remain to be investigated. The mechanisms engaging Notch in Th2 cell differentiation are summarized in Fig. 3.

4 Notch in Th17 Cell Differentiation

In addition to the well-documented role of Notch in Th1 function and Th2 cell differentiation, several reports have recently indicated that Notch signaling might be also involved in the generation of Th17 and Treg cells.

4.1 Notch Ligands Involved in CD4$^+$ Th17 Cell Differentiation

Pathogen-associated signals were shown to induce Dll4 expression on DCs, as discussed for Th1 differentiation. In addition, during in vitro Th17 differentiation conditions, Dll4 was reported to regulate Th17 cell activation: overexpression of Dll4 on DCs increased IL-17 secretion while neutralization decreased it (Mukherjee et al. 2009; Ito et al. 2009). Furthermore, an impact of Dll4 on Th17 activation was reported in vivo. Neutralization of Dll4 with an anti-Dll4 mAb during the development of *Mycobacterium*-induced pulmonary granulomas resulted in the formation of larger granulomas and secretion of decreased levels of Th17 cytokines (Ito et al. 2009). Dll4 induction was linked in this model to the presence of functional TLR9 signaling. In addition to the involvement of Dll4 in the activation of Th17 cells, the induction of Jagged1 mRNA by curdlan, a polysaccharide derived from *Candia albicans,* in human monocyte-derived DCs was shown to increase secretion of Th17 cell cytokines in vitro, and neutralization of Jagged 1 decreased curdlan-induced IL-17 secretion by T cells (Higashi et al. 2010).

4.2 Mechanisms of Notch Action During Th17 Cell Differentiation

Inhibition of Notch signaling using GSI during murine and human Th17 cell polarization in vitro resulted in reduced Th17-associated cytokine production (Keerthivasan et al. 2011). GSI-mediated inhibition of Notch signaling in vivo, in EAE, a model that promotes the development of Th17 cells, delayed disease progression as well as the severity of EAE symptoms (Minter et al. 2005; Jurynczyk et al. 2008; Mukherjee et al. 2009). Moreover, reduced levels of IL-17 produced by restimulated CD4$^+$ T cells from EAE-induced mice were observed following GSI treatment (Keerthivasan et al. 2011). The mode of action of GSI is not absolutely selective for Notch signaling but similar results obtained following neutralization of

Dll1 during Th17 polarization suggest a Notch-specific impact during Th17 cell differentiation (Ito et al. 2009). Neutralization of Dll1 resulted also in reduced EAE clinical score, while treatment with Dll1-Fc fusion protein worsened disease progression, correlating with an enhanced IFNγ secretion (Elyaman et al. 2007). These data were initially interpreted as an effect of Notch signaling on Th1 cell differentiation. However, during the development of EAE, the plasticity of Th17 cells has been demonstrated, and Th17 cells can shut off secretion of IL-17 to secrete IFNγ (Hirota et al. 2011). These data suggest that Notch signaling in EAE may result in a direct impact on Th17 differentiation or on the transition of classical Th17 cells to IFNγ-secreting cells. Whether the mechanism of Notch in these cells is identical to that observed in classical $CD4^+$ Th1 cells remains to be determined.

Further insights on the role of Notch in the differentiation of Th17 cells came from studies identifying consensus RBP-jκ binding sites on the *Il17* and *Rorc* promoters, suggesting that Notch may regulate IL-17 transcription after Dll4 activation (Keerthivasan et al. 2011; Mukherjee et al. 2009). However, confirmation of this mechanism in vivo will require further work. rDll4 was further shown to increase *Rorc* mRNA expression in T cells during Th17 differentiation in vitro, an effect that was lost in presence of GSI (Mukherjee et al. 2009).

Altogether these data reveal an impact of Notch signaling in Th17 differentiation. However, unlike what is observed for IL-4, inactivation of Notch signaling allows the differentiation of Th17 cells but has an impact on the level of IL-17 secretion, suggesting that Notch regulates Th17 function rather than differentiation.

5 The Involvement of Notch in the Differentiation of Treg Cells

5.1 Notch Ligands Inducing iTreg Cell Differentiation

The presence of Notch ligands, mostly Jagged ligands, was shown to induce antigen-specific iTreg differentiation and function in vitro in mice and humans (Hoyne et al. 2000). Overexpression of Jagged1 by human B lymphocytes was demonstrated to induce IL-10-secreting Treg cells that inhibited both proliferation and cytotoxic effector functions of autologous T cells in vitro (Vigouroux et al. 2003). In vivo, the adoptive transfer of Jagged2-expressing hematopoietic progenitor cells (HPCs) in NOD mice promoted the expansion of peripheral iTreg cells through activation of Notch3, an effect that prevented the development of spontaneous diabetes (Kared et al. 2006). In contrast to Jagged1 and Jagged2 that promote Treg cell differentiation, Dll4 was reported to suppress iTreg cell induction by TGF-β in vitro by inhibiting JAK3/STAT5 activation that is involved in Foxp3 expression, the master transcription regulator of Treg cells. Conversely, Dll4 blockade during the onset of EAE increased the number of Treg cells in the periphery and in the CNS, decreasing EAE severity score (Bassil et al. 2011). Thus most experiments point to a preponderant role of Jagged ligands in driving the induction of different types of Tregs.

5.2 *Notch1 and Notch3 are Involved in Treg Cell Differentiation*

Notch1 was demonstrated to cooperate with TGF-β signaling components (P-Smad3) in the activation of Foxp3 expression, suggesting a role of Notch1 in the maintenance of peripheral induced iTregs (Asano et al. 2008; Samon et al. 2008). The finding that CD4$^+$CD25$^+$ splenic Treg cells expressed higher amounts of Notch3 than CD4$^+$CD25$^-$ peripheral T cells indicated that Notch3 could be involved in the differentiation of iTreg cells (Anastasi et al. 2003). Overexpression of activated N3ICD in thymocytes and T cells in transgenic mice (N3-ICD-tg) induced up-regulation of CD25 in thymocytes, leading to accumulation of CD4$^+$ CD25$^+$ Treg cells in the thymus and peripheral lymphoid organs (Bellavia et al. 2000; Anastasi et al. 2003). This increased number of Treg cells in N3-ICD-tg mice was reported to play a protective role in experimental induced autoimmune diabetes (Anastasi et al. 2003). Notch3 cooperates with pTα in driving the increase of Foxp3-expressing Treg population, as well as the improvement of in vivo activity of naturally occurring Tregs (Campese et al. 2009). Furthermore, Notch3 was reported to be involved with protein kinase Cθ and p65/NF-κB subunit in order to modulate Foxp3 expression (Barbarulo et al. 2011). These data suggest that Notch1 and Notch3 could be involved in homeostasis and function of iTregs, possibly through a RBP-jκ-independent pathway. In addition, GSI-mediated inhibition of Notch reduced TGF-β-induced Foxp3 expression in Treg cells, affecting the suppressive activity of these cells in vitro and leading to a spontaneous lymphocyte infiltration of liver due to a defective Treg cell maintenance in vivo (Samon et al. 2008).

There are several lines of evidence suggesting the involvement of Notch signaling in Treg cell differentiation. It appears that Notch signaling is involved in the regulation of peripheral induced Treg cells. In addition, overexpression studies showed that Notch signaling could be involved in the differentiation of Treg cells in the thymus (natural Treg); however whether this occurs under physiological conditions remains to be determined.

6 Conclusions

Increasing data support a role for Notch signaling in the differentiation and/or function of distinct subsets of CD4$^+$ T helper cells. It is important to separate the effect of Notch on CD4$^+$ T helper differentiation from that on CD4$^+$ T helper functions. A striking example is given for CD4$^+$ Th1 cells that, in absence of Notch receptor expression, can differentiate into so called "competent" Th1 cells (Mohrs et al. 2005). These cells express T-bet, the master regulator of Th1 cells, and IFNγ, but fail to secrete IFNγ in response to a pathogen in vivo (Auderset et al. 2012). It will be interesting to unravel the mechanism involved in this control and to investigate whether Notch signaling regulates the release of other cytokines in a similar way.

The strongest effect of Notch on T helper differentiation is observed in Th2 cells, with a well-documented effect on the regulation of IL-4 transcription. However, one has to distinguish the direct effect of Notch signaling on the IL-4 promoter from that on a context-dependent usage of GATA3 exon 1a versus GATA3 exon 1b. As IL-4 is a cytokine important in driving Th2 differentiation, impaired secretion of IL-4 by Th2 cells will prevent the development of Th2 cells in most cases. However, Th2 cells can develop in vivo in an IL-4-independent way (Ansel et al. 2006; Finkelman et al. 2000; Jankovic et al. 2000). Thus, Notch signaling implication in the differentiation of Th2 cells may be either direct with a synergy between GATA3 exon 1a and IL-4 signaling, or indirect when GATA3 exon 1b is used and Th2 cell differentiation is IL-4-independent.

Convincing data show that the induction of specific Notch ligands by different pathogens is playing an essential role in triggering the development of functional T helper cells, modulating the secretion of cytokines secreted by T helper cells to different extents depending on the Th cell subset. How the expression of either Dll or Jagged ligands induce distinct T helper differentiation signals using similar Notch receptors remains to be determined.

It will also be interesting to investigate whether Notch has an impact in the differentiation of more recently described subsets such as follicular T helper cells (Crotty 2011) or Th9 cells (Veldhoen et al. 2008; Dardalhon et al. 2008). However, as observed for Th17 and Treg cells, plasticity in cytokine production by these latter Th subsets has been reported and may complicate interpretations in vivo.

Notch can act on $CD4^+$ Th cells in a canonical (Th2) (Amsen et al. 2004; Tanigaki et al. 2004; Tu et al. 2005) or noncanonical (RBP-jκ-independent) pathway (Th1) (Tu et al. 2005; Auderset et al. 2012), explaining some of the discrepancies observed. Determining the type of pathway involved in the function of Th cells will be particularly important to determine the use of inhibitors or activators of the Notch pathway that can be used to treat the many pathologies associated with the presence of distinct Th subsets.

Acknowledgments We would like to thank Dr. H. Robson MacDonald for discussions and critical reading of this chapter. This work was supported by grant 310030-129852/1 from the Swiss National Science Foundation and the ISREC foundation.

References

Amsen D, Blander JM, Lee GR, Tanigaki K, Honjo T, Flavell RA (2004) Instruction of distinct CD4 T helper cell fates by different notch ligands on antigen-presenting cells. Cell 117(4):515–526

Amsen D, Antov A, Jankovic D, Sher A, Radtke F, Souabni A, Busslinger M, McCright B, Gridley T, Flavell RA (2007) Direct regulation of Gata3 expression determines the T helper differentiation potential of Notch. Immunity 27(1):89–99

Amsen D, Antov A, Flavell RA (2009) The different faces of Notch in T-helper-cell differentiation. Nat Rev Immunol 9(2):116–124

Anastasi E, Campese AF, Bellavia D, Bulotta A, Balestri A, Pascucci M, Checquolo S, Gradini R, Lendahl U, Frati L, Gulino A, Di Mario U, Screpanti I (2003) Expression of activated Notch3 in transgenic mice enhances generation of T regulatory cells and protects against experimental autoimmune diabetes. J Immunol 171(9):4504–4511

Ansel KM, Djuretic I, Tanasa B, Rao A (2006) Regulation of Th2 differentiation and Il4 locus accessibility. Annu Rev Immunol 24:607–656. doi:10.1146/annurev.immunol.23.021704.115821

Asano N, Watanabe T, Kitani A, Fuss IJ, Strober W (2008) Notch1 signaling and regulatory T cell function. J Immunol 180(5):2796–2804

Asnagli H, Afkarian M, Murphy KM (2002) Cutting edge: Identification of an alternative GATA-3 promoter directing tissue-specific gene expression in mouse and human. J Immunol 168(9):4268–4271

Auderset F, Schuster S, Coutaz M, Koch U, Desgranges F, Merck E, MacDonald HR, Radtke F, Tacchini-Cottier F (2012) Redundant Notch1 and Notch2 signaling is necessary for IFNg secretion by T helper 1 cells during infection with Leishmania major. PLoS Pathog. doi:10.1371/journal.ppat.1002560

Barbarulo A, Grazioli P, Campese AF, Bellavia D, Di Mario G, Pelullo M, Ciuffetta A, Colantoni S, Vacca A, Frati L, Gulino A, Felli MP, Screpanti I (2011) Notch3 and canonical NF-kappaB signaling pathways cooperatively regulate Foxp3 transcription. J Immunol 186(11):6199–6206. doi:10.4049/jimmunol.1002136

Bassil R, Zhu B, Lahoud Y, Riella LV, Yagita H, Elyaman W, Khoury SJ (2011) Notch ligand delta-like 4 blockade alleviates experimental autoimmune encephalomyelitis by promoting regulatory T cell development. J Immunol 187(5):2322–2328. doi:10.4049/jimmunol.1100725

Bellavia D, Campese AF, Alesse E, Vacca A, Felli MP, Balestri A, Stoppacciaro A, Tiveron C, Tatangelo L, Giovarelli M, Gaetano C, Ruco L, Hoffman ES, Hayday AC, Lendahl U, Frati L, Gulino A, Screpanti I (2000) Constitutive activation of NF-kappaB and T-cell leukemia/lymphoma in Notch3 transgenic mice. Embo J 19(13):3337–3348. doi:10.1093/emboj/19.13.3337

Campbell DJ, Koch MA (2011) Phenotypical and functional specialization of FOXP3+ regulatory T cells. Nat Rev Immunol 11(2):119–130. doi:10.1038/nri2916

Campese AF, Grazioli P, Colantoni S, Anastasi E, Mecarozzi M, Checquolo S, De Luca G, Bellavia D, Frati L, Gulino A, Screpanti I (2009) Notch3 and pTalpha/pre-TCR sustain the in vivo function of naturally occurring regulatory T cells. Int Immunol 21(6):727–743. doi:10.1093/intimm/dxp042

Crotty S (2011) Follicular helper CD4 T cells (TFH). Annu Rev Immunol 29:621–663. doi:10.1146/annurev-immunol-031210-101400

Dardalhon V, Awasthi A, Kwon H, Galileos G, Gao W, Sobel RA, Mitsdoerffer M, Strom TB, Elyaman W, Ho IC, Khoury S, Oukka M, Kuchroo VK (2008) IL-4 inhibits TGF-beta-induced Foxp3+ T cells and, together with TGF-beta, generates IL-9+ IL-10+ Foxp3(-) effector T cells. Nat Immunol 9(12):1347–1355. doi:10.1038/ni.1677

Elyaman W, Bradshaw EM, Wang Y, Oukka M, Kivisakk P, Chiba S, Yagita H, Khoury SJ (2007) JAGGED1 and delta1 differentially regulate the outcome of experimental autoimmune encephalomyelitis. J Immunol 179 (9):5990–5998, 179/9/5990 [pii]

Fang TC, Yashiro-Ohtani Y, Del Bianco C, Knoblock DM, Blacklow SC, Pear WS (2007) Notch directly regulates Gata3 expression during T helper 2 cell differentiation. Immunity 27(1):100–110

Finkelman FD, Morris SC, Orekhova T, Mori M, Donaldson D, Reiner SL, Reilly NL, Schopf L, Urban JF Jr (2000) Stat6 regulation of in vivo IL-4 responses. J Immunol 164(5):2303–2310

Higashi T, Hashimoto K, Takagi R, Mizuno Y, Okazaki Y, Tanaka Y, Matsushita S (2010) Curdlan induces DC-mediated Th17 polarization via Jagged1 activation in human dendritic cells. Allergol Int 59(2):161–166. doi:10.2332/allergolint.09-OA-0103

Hirota K, Duarte JH, Veldhoen M, Hornsby E, Li Y, Cua DJ, Ahlfors H, Wilhelm C, Tolaini M, Menzel U, Garefalaki A, Potocnik AJ, Stockinger B (2011) Fate mapping of IL-17-producing T cells in inflammatory responses. Nat Immunol 12(3):255–263. doi:10.1038/ni.1993

Hoyne GF, Le Roux I, Corsin-Jimenez M, Tan K, Dunne J, Forsyth LM, Dallman MJ, Owen MJ, Ish-Horowicz D, Lamb JR (2000) Serrate1-induced notch signalling regulates the decision between immunity and tolerance made by peripheral CD4(+) T cells. Int Immunol 12(2):177–185

Ito T, Schaller M, Hogaboam CM, Standiford TJ, Sandor M, Lukacs NW, Chensue SW, Kunkel SL (2009) TLR9 regulates the mycobacteria-elicited pulmonary granulomatous immune response in mice through DC-derived Notch ligand delta-like 4. J Clin Invest 119(1):33–46. doi:10.1172/JCI35647

Ito T, Allen RM, Carson WFt, Schaller M, Cavassani KA, Hogaboam CM, Lukacs NW, Matsukawa A, Kunkel SL (2011) The critical role of Notch ligand delta-like 1 in the pathogenesis of influenza A virus (H1N1) infection. PLoS Pathog 7 (11):e1002341. doi:10.1371/journal.ppat.1002341

Jankovic D, Kullberg MC, Noben-Trauth N, Caspar P, Paul WE, Sher A (2000) Single cell analysis reveals that IL-4 receptor/Stat6 signaling is not required for the in vivo or in vitro development of CD4+ lymphocytes with a Th2 cytokine profile. J Immunol 164(6):3047–3055

Jurynczyk M, Jurewicz A, Bielecki B, Raine CS, Selmaj K (2008) Overcoming failure to repair demyelination in EAE: gamma-secretase inhibition of Notch signaling. J Neurol Sci 265(1–2): 5–11. doi:10.1016/j.jns.2007.09.007

Kang JH, Kim BS, Uhm TG, Lee SH, Lee GR, Park CS, Chung IY (2009) Gamma-secretase inhibitor reduces allergic pulmonary inflammation by modulating Th1 and Th2 responses. Am J Respir Crit Care Med 179(10):875–882

Kared H, Adle-Biassette H, Fois E, Masson A, Bach JF, Chatenoud L, Schneider E, Zavala F (2006) Jagged2-expressing hematopoietic progenitors promote regulatory T cell expansion in the periphery through notch signaling. Immunity 25(5):823–834. doi:10.1016/j.immuni.2006.09.008

Keerthivasan S, Suleiman R, Lawlor R, Roderick J, Bates T, Minter L, Anguita J, Juncadella I, Nickoloff BJ, Le Poole IC, Miele L, Osborne BA (2011) Notch signaling regulates mouse and human Th17 differentiation. J Immunol 187(2):692–701. doi:10.4049/jimmunol.1003658

Krawczyk CM, Sun J, Pearce EJ (2008) Th2 differentiation is unaffected by Jagged2 expression on dendritic cells. J Immunol 180 (12):7931–7937, 180/12/7931 [pii]

Krishnamoorthy N, Oriss TB, Paglia M, Fei M, Yarlagadda M, Vanhaesebroeck B, Ray A, Ray P (2008) Activation of c-Kit in dendritic cells regulates T helper cell differentiation and allergic asthma. Nat Med 14(5):565–573. doi:10.1038/nm1766

Liotta F, Frosali F, Querci V, Mantei A, Fili L, Maggi L, Mazzinghi B, Angeli R, Ronconi E, Santarlasci V, Biagioli T, Lasagni L, Ballerini C, Parronchi P, Scheffold A, Cosmi L, Maggi E, Romagnani S, Annunziato F (2008) Human immature myeloid dendritic cells trigger a TH2-polarizing program via Jagged-1/Notch interaction. J Allergy Clin Immunol 121 (4):1000–1005 e1008. doi:10.1016/j.jaci.2008.01.004

Locksley RM (2009) Nine lives: plasticity among T helper cell subsets. J Exp Med 206(8): 1643–1646. doi:10.1084/jem.20091442

Maekawa Y, Tsukumo S, Chiba S, Hirai H, Hayashi Y, Okada H, Kishihara K, Yasutomo K (2003) Delta1-Notch3 interactions bias the functional differentiation of activated CD4+ T cells. Immunity 19(4):549–559

Minter LM, Turley DM, Das P, Shin HM, Joshi I, Lawlor RG, Cho OH, Palaga T, Gottipati S, Telfer JC, Kostura L, Fauq AH, Simpson K, Such KA, Miele L, Golde TE, Miller SD, Osborne BA (2005) Inhibitors of gamma-secretase block in vivo and in vitro T helper type 1 polarization by preventing Notch upregulation of Tbx21. Nat Immunol 6(7):680–688

Mohrs K, Wakil AE, Killeen N, Locksley RM, Mohrs M (2005) A two-step process for cytokine production revealed by IL-4 dual-reporter mice. Immunity 23(4):419–429. doi:10.1016/j.immuni.2005.09.006

Mukherjee S, Schaller MA, Neupane R, Kunkel SL, Lukacs NW (2009) Regulation of T cell activation by Notch ligand, DLL4, promotes IL-17 production and Rorc activation. J Immunol 182(12):7381–7388. doi:10.4049/jimmunol.0804322

Napolitani G, Rinaldi A, Bertoni F, Sallusto F, Lanzavecchia A (2005) Selected Toll-like receptor agonist combinations synergistically trigger a T helper type 1-polarizing program in dendritic cells. Nat Immunol 6(8):769–776. doi:10.1038/ni1223

Okamoto M, Takeda K, Joetham A, Ohnishi H, Matsuda H, Swasey CH, Swanson BJ, Yasutomo K, Dakhama A, Gelfand EW (2008) Essential role of Notch signaling in effector memory CD8+ T cell-mediated airway hyperresponsiveness and inflammation. J Exp Med 205(5):1087–1097. doi:10.1084/jem.20072200

Okamoto M, Matsuda H, Joetham A, Lucas JJ, Domenico J, Yasutomo K, Takeda K, Gelfand EW (2009) Jagged1 on dendritic cells and Notch on CD4+ T cells initiate lung allergic responsiveness by inducing IL-4 production. J Immunol 183(5):2995–3003. doi:10.4049/jimmunol.0900692

Ong CT, Sedy JR, Murphy KM, Kopan R (2008) Notch and presenilin regulate cellular expansion and cytokine secretion but cannot instruct Th1/Th2 fate acquisition. PLoS One 3(7):e2823

O'Shea JJ, Paul WE (2010) Mechanisms underlying lineage commitment and plasticity of helper CD4+ T cells. Science 327 (5969):1098-1102. doi:10.1126/science.1178334, 327/5969/1098 [pii]

Osipo C, Golde TE, Osborne BA, Miele LA (2008) Off the beaten pathway: the complex cross talk between Notch and NF-kappaB. Lab Invest 88 (1):11–17. doi:10.1038/labinvest.3700700, 3700700 [pii]

Radtke F, Fasnacht N, Macdonald HR (2010) Notch signaling in the immune system. Immunity 32(1):14–27

Rudd BD, Schaller MA, Smit JJ, Kunkel SL, Neupane R, Kelley L, Berlin AA, Lukacs NW (2007) MyD88-mediated instructive signals in dendritic cells regulate pulmonary immune responses during respiratory virus infection. J Immunol 178(9):5820–5827

Samon JB, Champhekar A, Minter LM, Telfer JC, Miele L, Fauq A, Das P, Golde TE, Osborne BA (2008) Notch1 and TGFbeta1 cooperatively regulate Foxp3 expression and the maintenance of peripheral regulatory T cells. Blood 112(5):1813–1821. doi:10.1182/blood-2008-03-144980

Sanalkumar R, Dhanesh SB, James J (2010) Non-canonical activation of Notch signaling/target genes in vertebrates. Cell Mol Life Sci 67(17):2957–2968. doi:10.1007/s00018-010-0391-x

Sandy AR, Maillard I (2009) Notch signaling in the hematopoietic system. Expert Opin Biol Ther 9(11):1383–1398

Schaller MA, Neupane R, Rudd BD, Kunkel SL, Kallal LE, Lincoln P, Lowe JB, Man Y, Lukacs NW (2007) Notch ligand Delta-like 4 regulates disease pathogenesis during respiratory viral infections by modulating Th2 cytokines. J Exp Med 204(12):2925–2934. doi:10.1084/jem.20070661

Shi J, Fallahi M, Luo JL, Petrie HT (2011) Nonoverlapping functions for Notch1 and Notch3 during murine steady-state thymic lymphopoiesis. Blood 118(9):2511–2519. doi:10.1182/blood-2011-04-346726

Shin HM, Minter LM, Cho OH, Gottipati S, Fauq AH, Golde TE, Sonenshein GE, Osborne BA (2006) Notch1 augments NF-kappaB activity by facilitating its nuclear retention. Embo J 25(1):129–138

Skokos D, Nussenzweig MC (2007) CD8- DCs induce IL-12-independent Th1 differentiation through delta 4 Notch-like ligand in response to bacterial LPS. J Exp Med 204 (7):1525–1531. doi:10.1084/jem.20062305, jem.20062305 [pii]

Sun J, Krawczyk CJ, Pearce EJ (2008) Suppression of Th2 cell development by Notch ligands Delta1 and Delta4. J Immunol 180 (3):1655–1661, 180/3/1655 [pii]

Tacchini-Cottier F, Allenbach C, Otten LA, Radtke F (2004) Notch1 expression on T cells is not required for CD4+ T helper differentiation. Eur J Immunol 34(6):1588–1596

Tanaka S, Tsukada J, Suzuki W, Hayashi K, Tanigaki K, Tsuji M, Inoue H, Honjo T, Kubo M (2006) The interleukin-4 enhancer CNS-2 is regulated by Notch signals and controls initial expression in NKT cells and memory-type CD4 T cells. Immunity 24 (6):689–701. doi:10.1016/j.immuni.2006.04.009, S1074-7613(06)00267-6 [pii]

Tanigaki K, Tsuji M, Yamamoto N, Han H, Tsukada J, Inoue H, Kubo M, Honjo T (2004) Regulation of alphabeta/gammadelta T cell lineage commitment and peripheral T cell responses by Notch/RBP-J signaling. Immunity 20 (5):611–622, S1074761304001098 [pii]

Tu L, Fang TC, Artis D, Shestova O, Pross SE, Maillard I, Pear WS (2005) Notch signaling is an important regulator of type 2 immunity. J Exp Med 202(8):1037–1042

van den Brandt J, Kwon SH, McPherson KG, Petrovic S, Zettl A, Muller-Hermelink HK, Reichardt HM (2006) Unexpected features of acute T lymphoblastic lymphomas in Notch1IC transgenic rats. Eur J Immunol 36(8):2223–2234. doi:10.1002/eji.200535791

Veldhoen M, Uyttenhove C, van Snick J, Helmby H, Westendorf A, Buer J, Martin B, Wilhelm C, Stockinger B (2008) Transforming growth factor-beta 'reprograms' the differentiation of T helper 2 cells and promotes an interleukin 9-producing subset. Nat Immunol 9(12):1341–1346. doi:10.1038/ni.1659

Vigouroux S, Yvon E, Wagner HJ, Biagi E, Dotti G, Sili U, Lira C, Rooney CM, Brenner MK (2003) Induction of antigen-specific regulatory T cells following overexpression of a Notch ligand by human B lymphocytes. J Virol 77(20):10872–10880

Worsley AG, LeibundGut-Landmann S, Slack E, Phng LK, Gerhardt H, Reis e Sousa C, MacDonald AS (2008) Dendritic cell expression of the Notch ligand jagged2 is not essential for Th2 response induction in vivo. Eur J Immunol 38(4):1043–1049

Zheng W, Flavell RA (1997) The transcription factor GATA-3 is necessary and sufficient for Th2 cytokine gene expression in CD4 T cells. Cell 89(4):587–596

Notch Signaling in Alloreactive T Cell Immunity

Jooho Chung and Ivan Maillard

Abstract Alloreactive T cell immunity mediates the recognition of foreign tissue antigens in recipients of organ transplants. After solid organ transplantation, activation of host T cells by donor alloantigens can trigger rejection of the implanted organ. Global life-long immunosuppression is necessary to prevent or to minimize organ rejection. After bone marrow or hematopoietic cell transplantation (allo-BMT), donor-derived T cells recognize host alloantigens, inducing both beneficial graft-versus-tumor (GVT) effects as well as detrimental graft-versus-host disease (GVHD). Preventing GVHD without eliminating GVT activity is an essential goal to maximize the safety and efficacy of allo-BMT. In this review, we discuss emerging findings that have identified the Notch pathway as a central player in the regulation of T cell alloimmunity. In view of these effects, Notch signaling in T cells should be considered as an attractive new therapeutic target to achieve beneficial immunomodulation following allo-BMT and other types of allogeneic transplantation.

J. Chung · I. Maillard (✉)
Life Sciences Institute,
e-mail: imaillar@umich.edu

J. Chung · I. Maillard
Cellular and Molecular Biology Program,

I. Maillard
Department of Medicine, Division of Hematology-Oncology,

I. Maillard
Department of Cell and Developmental Biology,
University of Michigan, Ann Arbor, 48104, USA

Current Topics in Microbiology and Immunology (2012) 360: 135–150
DOI: 10.1007/82_2012_226

Published Online: 12 June 2012

Contents

1 Introduction to Notch signaling in Mature T cells

Notch is a highly conserved cell–cell communication pathway with multiple functions in development and tissue homeostasis (Kopan and Ilagan 2009; Louvi and Artavanis-Tsakonas 2012). In the hematopoietic system, Notch was first identified for its oncogenic activity in T cell acute lymphoblastic leukemia (Ellisen et al. 1991; Weng et al. 2004). At physiological levels of signaling, Notch is required at early stages of T lineage development in the thymus (Radtke et al. 1999; Han et al. 2002; Sambandam et al. 2005; Tan et al. 2005). The effects of Notch are regulated in vivo through tight control of signaling intensity (Sandy and Maillard 2009;Radtke et al. 2010; Yuan et al. 2010). Lymphoid progenitors experience a sharp increase in Notch signaling upon entry into the thymus (Sambandam et al. 2005; Tan et al. 2005) as a result of their exposure to a high density of the Notch ligand Delta-like-4 in the thymic epithelium (Koch et al. 2008). High levels of Notch signaling are maintained until progenitors successfully cross the pre-T cell receptor checkpoint, after which signaling intensity is rapidly downregulated (Ciofani et al. 2004;Ciofani and Zuniga-Pflucker 2005; Tan et al. 2005; Maillard et al. 2006; Yashiro-ohtani et al. 2009). Downstream of this checkpoint, $CD4^+CD8^+$ double positive (DP) thymocytes experience little if any Notch signaling when undergoing positive and negative selection. Thus, unlike forced induction of Notch signaling, Notch blockade in DP thymocytes does not interfere with T cell development (Robey et al. 1996; Wolfer et al. 2001; Fowlkes and Robey 2002; Tanigaki et al. 2004; Tu et al. 2005).

In addition to the effects of Notch signaling in T cell development, increasing attention is being devoted to its role in peripheral T cell immunity (reviewed in (Maillard et al. 2005; Osborne and Minter 2007; Amsen et al. 2009; Radtke et al. 2010; Yuan et al. 2010)). Mature T cells predominantly express Notch1 and Notch2 receptors, which can engage Notch ligands during immune responses (Amsen et al. 2007; Fiorini et al. 2009; Auderset et al. 2012). Innate signals regulate expression of Notch ligands of the Jagged (Jagged1,2) and Delta-like (Dll1, 3, 4) families. Toll-like receptor-mediated signals induce expression of Delta-like ligands in dendritic cells (DCs) and other antigen-presenting cells (APCs) (Amsen et al. 2004; Napolitani et al. 2005; Schaller et al. 2007; Ito et al.

2009). Additional sources of Notch ligands may be available to T cells in tissues (e.g. from stromal cells, vascular structures or epithelial elements). Altogether, this sets the stage for a highly regulated context-dependent exposure of T cells to Notch signaling in different immune responses, with specific roles for individual Notch ligands and receptors.

Complementary gain-of-function and loss-of-function experiments have been used extensively to study Notch in lower organisms and mammals. Although this approach has often proven to be fruitful, caution is required when Notch signaling intensity is tightly regulated. For example, Notch stimulation and Notch blockade may not have opposite effects in cells experiencing a low intensity of Notch signaling, as shown in DP thymocytes or in adult hematopoietic stem cells (Robey et al. 1996; Varnum-Finney et al. 2000; Wolfer et al. 2001; Varnum-Finney et al. 2003; Tanigaki et al. 2004; Maillard et al. 2008; Gao et al. 2009; Delaney et al. 2010). In addition, forced exposure to Notch ligands or forced signaling through individual Notch receptors may have functional consequences that do not represent the in vivo function of the pathway. These considerations apply to the study of Notch in mature T cells. Although in vitro studies and gain-of-function experiments can generate useful working hypotheses, we believe that optimal experimental systems should involve stringent Notch loss-of-function systems applied in vivo. In this review, we discuss an emerging role for Notch signaling in T cell alloimmunity in light of these considerations.

2 Introduction to Alloreactive T cell Immunity

Allogeneic T cell responses drive immune reactivity against foreign tissue antigens. Alloimmunity plays an essential role in several situations of high medical significance. Host alloreactive T cells mediate immune rejection of implanted organs after solid organ transplantation (e.g. heart, lung, liver, or kidney transplantation). With up to 100,000 organ transplants performed each year worldwide, T cell-mediated immune rejection represents a sizable public health problem. Current medical practice involves administration of life-long global immunosuppression (e.g. calcineurin inhibitors) to prevent organ rejection and corticosteroids to treat breakthrough rejection. This strategy is more successful for acute than for chronic rejection. In addition, life-long immunosuppression is associated with high costs and side effects, including increased risk of opportunistic infections and secondary malignancies. As an alternative, researchers have investigated strategies to induce host tolerance to the implanted organ, for example by interference with costimulatory molecules (Li et al. 2009) or with non-myeloablative hematopoietic cell transplantation (Kawai et al. 2008; Leventhal et al. 2012). At this stage, these strategies have not translated into treatments that can be used in routine clinical practice.

After allogeneic bone marrow or hematopoietic cell transplantation (allo-BMT), rejection is rare due to the immunosuppressive effects of the conditioning regimen on the recipient (Welniak et al. 2007). However, mature donor-derived T

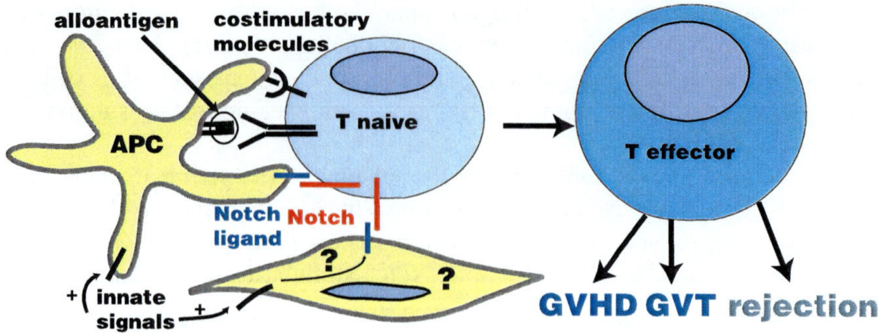

Fig. 1 Regulation of alloreactive T cell activation and differentiation. Signals from the innate immune system activate antigen-presenting cells (APCs) and enable them to present alloantigens to na T cells in the presence of costimulatory signals and/or Notch signaling. Notch ligands are provided by APCs or other cell types (noted by a question mark). Upon activation, na T cells become effector T cells, and mediate graft rejection, graft-versus-host disease, and/or graft-versus tumor effects

cells present in the transplant inoculum drive reactivity to alloantigens in host tissues and host-derived tumors (Shlomchik 2007; Welniak et al. 2007; Ferrara et al. 2009). Efficient donor T cell-mediated graft-versus-tumor (GVT) activity results in cancer cell killing (Wu and Ritz 2006; Welniak et al. 2007; Kolb et al. 2008; Falkenburg and Warren 2011). This is essential for the success of allo-BMT since the procedure is performed most often for patients with leukemia, lymphoma or other hematological malignancies (>75 % of the ~25,000 allo-BMTs performed worldwide each year). Unfortunately, GVT is often associated with reactivity of donor T cells against normal host tissues, leading to graft-versus-host disease (GVHD). GVHD is the most significant complication that limits the success and broad applicability of allo-BMT. Despite the universal use of potent immunosuppression in allo-BMT recipients, acute and chronic GVHD still occur in a substantial fraction of patients. Although intensifying the degree of global immunosuppression decreases the risk of GVHD, it comes at the price of impairing GVT efficiency. As a result, patients experience less GVHD but more tumor relapses, and thus no improvement in their overall survival (Ho and Soiffer 2001; Welniak et al. 2007). Therefore, progress in the field will require identification of new approaches to decrease GVHD severity without eliminating efficient GVT.

Alloantigens can be associated with the major histocompatibility complex (MHC), or independent of it ("minor histocompatibility antigens") (Shlomchik 2007). Priming and differentiation of alloreactive T cells occurs in several steps (Fig. 1) (Shlomchik 2007; Ferrara et al. 2009). Tissue damage and inflammation trigger activation and maturation of APCs through Toll-like receptor signaling and other innate pathways (Alegre and Chong 2009). In allo-BMT, toxic conditioning regimens including total body irradiation and/or chemotherapy contribute to APC activation and maturation. Activated APCs expressing costimulatory molecules prime na alloreactive T cells. In mouse allo-BMT models, host DCs play a dominant role in stimulating donor-derived alloreactive $CD8^+$ T cells rapidly after transplantation (Shlomchik et al. 1999; Zhang et al. 2002; Tawara et al. 2010). Donor APCs are important in certain settings as well (Anderson et al. 2005). Interestingly, recent work suggests that nonhematopoietic APCs also play a critical role in driving $CD4^+$ T cell (Jones et al. 2003; Koyama et al. 2012; Li et al. 2012; Toubai et al. 2012). Alloreactive T cells undergo activation, proliferation and expansion in lymphoid tissues. These cells then acquire effector functions through differentiation along various $CD4^+$ and $CD8^+$ T cell lineages. Much work has been performed to understand how effector pathways are regulated in alloreactive T cells, and this will be discussed later as it relates to Notch signaling in these cells. After priming in lymphoid organs, effector T cells undergo chemokine-directed migration into target tissues, mediating rejection (solid organ transplantation), as well as GVHD and GVT (allo-BMT) (Wysocki et al. 2005). Clinically important targets of GVHD include the skin, intestine, liver, lung, and thymus. Other immune cells collaborate with T cells in inducing tissue damage, especially in the chronic phase of the response. Activation of conventional T cells occurs simultaneously with activation and expansion of regulatory T cells, which play a critical role in the outcome of the alloreactive T cell response (Nguyen et al. 2007; 2007; Welniak et al. 2006; Coghill et al. 2008).

3 Overview of Published Work on Notch and Alloreactivity

In Table 1, we summarize all the published work that, to our knowledge, has investigated a role for Notch in alloimmunity. This body of work involves multiple experimental systems in mice (Wong et al. 2003; Fu et al. 2011; Riella et al. 2011; Toubai et al. 2011Zhang et al. 2011) and more limited in vitro studies with human T cells (Yvon et al. 2003). Very divergent methods of genetic or pharmacologic Notch modulation were used, including both gain-of-function and loss-of-function approaches in vivo or in vitro. In addition, some studies targeted signaling downstream of all Notch receptors and ligands (Fu et al. 2011; Zhang et al. 2011), while others focused on overexpression or blockade of individual ligands (Wong et al. 2003; Yvon et al. 2003; Riella et al. 2011). Although most studies focused on direct effects of Notch signaling on T cells, it is becoming apparent that Notch also exerts important functions in other immune cells, such as DCs (Caton et al. 2007; Bugeon et al. 2008; Lewis et al. 2011; Toubai et al. 2011). At this stage, there is no single model that can account for all the published observations listed in Table 1 in a coherent fashion. In fact, some of the findings appear contradictory. We believe that discrepancies may result from observations made in divergent experimental systems. As the field moves forward, clarification will emerge from the use of in vivo Notch loss-of-function systems targeting all Notch signals, or individual Notch ligands and receptors, in specific immune cell types.

Table 1 Overview of published work exploring the role of notch signaling in alloreactive T cell immunity

Alloreactivity model	Modulation of Notch signaling	Outcome	Suggested mechanism	Citation
Allograft rejection				
MHC-mismatched mixed lymphocyte reaction	Jagged1 overexpression in EBV-transformed lymphoblastoid cells	↓ proliferation ↓↓↓ cytotoxicity	Induction of regulatory T cells	Yvon et al., 2003
Heterotopic cardiac transplantation	IV delivery of Delta1-overexpressing L-cells transfected with MHC I/II	↑↑↑ graft survival	Induction of hyporesponsive $CD8^+$ T cells	Wong et al., 2003
Heterotopic cardiac transplantation in *Cd28*$^{-/-}$ or CTLA4-Ig	Neutralizing Dll1-specific monoclonal antibody	↑↑ graft survival ↓↓ Th1 cytokines ↑↑ Th2 cytokines	Skewing of Th1 response to Th2 response	Riella et al., 2011
Skin allograft; Heterotopic cardiac transplantation	*Rbpj*$^{f/f}$ x *Lck-Cre*	↓ graft survival ↑ Th1, Th2, Th17 cytokines ↑ proliferation ↓ regulatory T cell function	Loss of regulatory T cell function, enhanced cytokine production	Fu et al., 2011
Graft-versus-host disease				
B6 -> BALB/c B6 -> B6 x DBA/2 F1 B6 -> BALB/b*	*ROSA26*$^{DNMAML/f}$ x *Cd4-Cre* *Rbpj*$^{f/f}$ x *Cd4-Cre* *Notch1*$^{f/f}$;*Notch2*$^{f/f}$ x *Cd4-Cre**	↓↓↓ Th1, Th2, Th17 cytokines ↑ regulatory T cells ↔ cytotoxicity ↔ proliferation	Decreased cytokine production, increased regulatory T cells	Zhang et al., 2011, *unpublished data
B6 -> [Ik$^{-/-}$ -> B6Ly5.2] C3H.sw -> [Ik-/- -> B6Ly5.2]	Ik$^{-/-}$ leading to ↑ Notch signaling in DCs	↑↑ Th1, Th17 cytokines ↔ regulatory T cells	Increased cytokine production	Toubai et al., 2011

4 Notch Signaling in Tolerance and Organ Rejection

Interest in a role for Notch signaling in tolerance induction and immunity was sparked by observations from Lamb and coworkers in an experimental system that did not involve alloantigens, but a peptide antigen from the house dust mite protein Der p1 (Hoyne et al. 2000). These investigators engineered mouse DCs to overexpress the Notch ligand Jagged1, a homolog of the fly Serrate (DC-J1). DC-J1 were pulsed with antigen and adoptively transferred into naive mice. Mice exposed to DC-J1 failed to respond productively to a subsequent immunization with peptide and adjuvant in an antigen-specific manner. Hyporesponsiveness was long-lived and could be transferred into na secondary mice via infusion of $CD4^+$ T cells. This suggested that exposure to DC-J1 had induced T cells with a regulatory function that could suppress an immune response to the model antigen.

As these observations were relevant to tolerance induction, the Brenner group tested the ability of Jagged1-overexpressing APCs to dampen T cell responses to alloantigen (Yvon et al. 2003). Epstein-Barr virus (EBV)-transformed lymphoblastoid B-cells (LCLs) were infected with a chimeric adenovirus encoding Jagged1, such that *Jagged1* transcripts were more than 60-fold higher than in control virus-infected cells. When compared to T cells cocultured with allogeneic control

LCL cells, na CD4$^+$ T-cells cocultured with allogeneic LCL cells overexpressing Jagged1 (LCL-J1) proliferated less, while na CD4$^+$ and CD8$^+$ T-cells cocultured with allogeneic LCL-J1 cells displayed reduced cytotoxic ability (Yvon et al. 2003). As in the Lamb study (Hoyne et al. 2000), the suppressive effects induced by LCL-J1 cells were transferrable into separate fresh cultures. Furthermore, supernatants from cocultures of CD4$^+$ and CD8$^+$ T-cells with allogeneic LCL-J1 cells contained less IL-2, IFN-γ, and IL-5, but unchanged levels of IL-10, when compared to cocultures with control LCL cells.

Interestingly, this group performed similar experiments using autologous T cells and EBV-transformed LCL-J1 cells (Vigouroux et al. 2003). In this system, antigen reactivity was directed to viral antigens rather than alloantigens. Defective T cell proliferation and anti-tumor cytotoxicity were observed upon exposure to LCL-J1 cells, just as in the allogeneic setting (Vigouroux et al. 2003; Yvon et al. 2003). However, J1 overexpression had a different effect on cytokine production, as T cells secreted increased levels of IL-10 but unchanged levels of IFN-γ, IL-5, and IL-2. These divergent effects, despite the use of nearly identical experimental systems, suggest that cytokine production may not be directly regulated by Notch. Jagged1 overexpression may have induced other conserved effects on T cell activation and function, with indirect context-dependent effects on cytokine secretion.

The dampening effects of Notch ligands on T cell function were corroborated by in vivo cardiac allograft rejection studies, in which mouse L cell fibroblasts were engineered to overexpress both Dll1 and MHC class I (K^b/Dll1) and/or MHC class II (A^b/Dll1) alloantigens (Wong et al. 2003). Intravenous delivery of both K^b/Dll1 and A^b/Dll1 cells, but not just K^b/Dll1 or A^b/Dll1 cells, prior to heterotopic transplantation of cardiac allografts resulted in a delay in rejection. Depletion of CD8$^+$ host cells, but not CD4$^+$ host cells, abrogated the protective effects of K^b/Dll1 plus A^b/Dll1 delivery. In vitro stimulation of T cells with recombinant Dll1 resulted in increased IL-10 production concomitant with decreases in IFN-γ and IL-2. These data suggested that Dll1 overexpression on APCs could act to induce tolerance to alloantigens and in turn tolerance to allografts. The experimental system did not allow for the authors to establish if Dll1 overexpression functioned entirely through its direct effects on the T cell compartment. Alternatively, Dll1 may have modulated the function of the L cells themselves, or that of other recipient cells, including endogenous APCs.

Taken together, aforementioned studies brought a pioneering observation to the alloimmunity field by demonstrating that forced exposure to J1 or Dll1 could induce a population of T cells with suppressed reactivity and enhanced regulatory function. However, the following considerations are important when assessing the overall significance of this work, especially in terms of the in vivo regulation of alloreactive T cell responses by Notch: 1) J1 and Dll1 were massively overexpressed in a unregulated fashion; 2) T cells were exposed to irradiated cells (LCL-J1) in vitro or heterologous cells in vivo (L cells); 3) the effect of ligand overexpression on APC function was not taken into consideration. Indeed, the recent literature has suggested that J1 itself can directly promote IL-10 and IL-2 production by DCs (Bugeon et al. 2008); 4) competition with endogenous Notch ligands may have occurred. Thus, the physiological role of Notch signaling in the regulation of alloreactive T cell responses cannot be predicted on the basis of these results.

In contrast to the previously mentioned gain-of-function studies, Riella et al. blocked Dll1 systemically with monoclonal antibodies at the time of cardiac allograft transplantation (Riella et al. 2011). These experiments yielded results that were contradictory to conclusions from the Dll1-overexpression studies [(Wong et al. 2003; Riella et al. 2011). While systemic administration of anti-Dll1 alone minimally delayed cardiac allograft rejection, it significantly prolonged allograft survival when provided in conjunction with costimulatory signal blockade, either through genetic deletion of *Cd28* or with CTLA4-Ig. Splenocytes from anti-Dll1 treated mice secreted decreased levels of IL-2, IL-6, and IL-17, while producing increased amounts of the Th2 cytokines IL-4 and IL-5. Genetic deletion of the Th2-related *Stat6* in recipient mice abrogated the protective effects of Dll1 blockade. These data suggested that the skewing of alloreactive T cell cytokine production toward Th2-related cytokines were responsible for the effects of Dll1 inhibition. This hypothesis was supported by other work demonstrating that Delta ligands promoted Th1 or blocked Th2 differentiation (Amsen et al. 2004; Schaller et al. 2007). The experimental approach, however, did not allow for the authors to determine if the protective effects of Dll1 blockade were achieved entirely through its effects on Notch signaling in T cells, or also via modulation of Notch signaling in other compartments, such as DCs. Furthermore, systemic administration of Dll1 blocking antibodies only minimally impaired marginal zone B (MZB) cell development in this report (Riella et al. 2011). As MZB cells depend on Dll1-mediated signals in a dosage-sensitive fashion (Hozumi et al. 2004), it is possible that systemic Dll1 blockade was only partial in this study, leading the authors to underestimate the effects of Dll1 blockade on allograft rejection.

Contrary to the Najafian group's findings that pharmacologic blockade of Dll1 prolonged allograft survival, Fu et al. found that Lck-Cre-mediated deletion of *Rbpjk* (encoding CSL/RBP-Jκ) in the T cell compartment of allograft recipients modestly accelerated both cardiac and skin allograft rejection (Fu et al. 2011). T cells lacking CSL/RBP-Jk are incapable of Notch-mediated transcriptional activation (Han et al. 2002). Only very small numbers of animals were reported. $CD4^+$ and $CD8^+$ T-cells from conditionally deleted *Rbpj* allograft recipient mice proliferated at a greater rate, and expressed increased levels of IFN-γ, IL-4, and IL-17 when compared to $CD4^+$ and $CD8^+$ cells from control allograft recipient mice, even without antigen-specific restimulation. Surprisingly, more than 30 % of $CD4^+$ T-cells from the peripheral blood of allograft recipient mice produced Th1 cytokines on the first day after transplantation. These data are inconsistent with previous reports demonstrating a much lower frequency of alloreactive T cells (Suchin et al. 2001) and with the kinetics of T cell priming after organ transplantation. Therefore, it is possible that the analysis included non-alloreactive T cells and those T cells were hyperreactive on baseline in these mice (which has not been observed in other genetic models of Notch blockade in mature T cells) (Wolfer et al. 2001; Tanigaki et al. 2004; Tu Let a 2005). Thus, this work may not allow definitive conclusions about the role of Notch signaling during graft rejection.

5 Notch Signaling in Graft-Versus-Host Disease

GVHD remains the most serious complication of allo-BMT, limiting its effectiveness and broad applicability. Donor-derived T cells mediate damage of GVHD target tissues through recognition of host-derived MHC or non-MHC alloantigens (Fig. 1). Despite universal prophylaxis, GVHD occurs in a large fraction of patients, and up to 50 % of patients with severe acute GVHD fail to respond to steroid-based treatment in a sustained fashion. In addition, donor T cell depletion or global immunosuppression increase the risk of infection and decrease the efficiency of graft-versus-tumor (GVT) effects, leading to an increased risk of tumor relapse. Thus, patients could markedly benefit from the identification of new molecular targets and from therapeutic strategies that decrease GVHD without compromising efficient GVT activity.

Our laboratory has discovered a new essential function of Notch signaling in the regulation of pathogenic T cells that mediate GVHD (Zhang et al. 2011). To assess the overall impact of Notch signals in mouse models of allo-BMT, we conditionally expressed a Dominant Negative Mastermind-like (DNMAML) pan-Notch inhibitor in donor T cells. DNMAML is a truncated fragment of the Mastermind-like1 (MAML1) coactivator fused to GFP, with potent dominant negative activity that blocks Notch-mediated transcriptional activation downstream of all four Notch receptors (Weng et al. 2003; Maillard et al. 2004; Sambandam et al. 2005; Tu et al. 2005; Maillard et al. 2006; Maillard et al. 2008). Conditional Cre-mediated DNMAML expression was achieved at the DP stage of T cell development in *Cd4-Cre* x *ROSA26*^{DNMAMLf} mice (Tu et al. 2005; Zhang et al. 2011). This allowed for efficient Notch signaling blockade in $CD4^+$ and $CD8^+$ T-cells without interference with T cell development or na T cell homeostasis. DNMAML expression in donor T cells resulted in markedly decreased secretion of inflammatory cytokines and near-complete protection from acute GVHD in multiple models of allo-BMT: a major antigen mismatched model (B6 – >BALB/c), a parent to F1 model (B6 – >BDF1) (Zhang et al. 2011), as well as a minor antigen mismatched model (B6 – >BALB/b) (unpublished data) (Fig. 2). Protection from GVHD was not mediated through global immunosuppression (Fig. 3), as Notch signaling-deficient alloreactive T cells proliferated and expanded in vivo as well, or even better, than wild-type alloreactive T cells. Mice receiving DNMAML T cells also showed an increased accumulation of regulatory T cells. Importantly, DNMAML T cells retained potent cytotoxic potential and GVT activity, as recipients of DNMAML T cells controlled a leukemia challenge, even with high numbers of cancer cells ((Zhang et al. 2011), unpublished data). This led to long-term survival free of both leukemia and severe GVHD. Altogether, our findings identify Notch signaling in donor T cells as an attractive therapeutic target to achieve beneficial immunomodulation after allo-BMT.

To understand the mechanisms of Notch action in alloreactive T cells, we are using several complementary Notch loss-of-function models. We observed similar effects upon DNMAML expression or *Rbpj* inactivation in donor T cells (Zhang

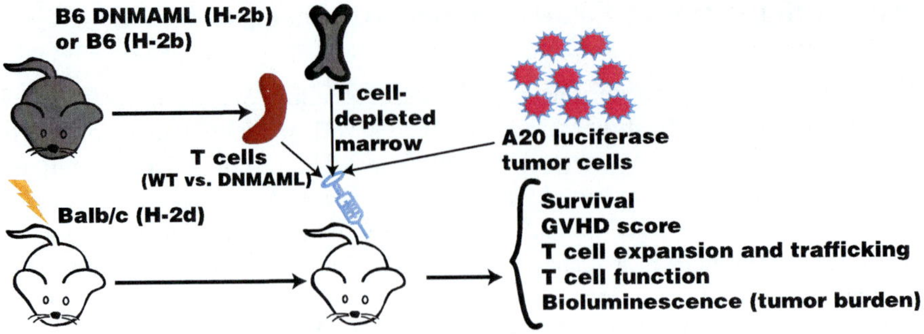

Fig. 2 Models of notch inhibition after allogeneic bone marrow transplantation. Donor T cells are isolated from B6 *Rosa26*^{DNMAMLf} x *Cd4-Cre* or B6 wild-type (WT) spleens and adoptively transferred with T cell depleted marrow and luciferase-expressing A20 tumor cells into lethally irradiated MHC-mismatched BALB/c recipient mice (B6 – >BALB/c) (Zhang et al. 2011). Recipient mice are assessed for the parameters listed. Similar experiments have been performed with T cells from *Notch1*$^{f/f}$;*Notch2*$^{f/f}$ x *Cd4-Cre* and *Rbpj*$^{f/f}$ x *Cd4-Cre* donor mice, with comparable results. Alternative models involving minor alloantigen mismatch (B6 – > BALB/b) and no irradiation (B6 – > BDF1) have also revealed an important role for Notch signaling [(Zhang et al. 2011), unpublished data]

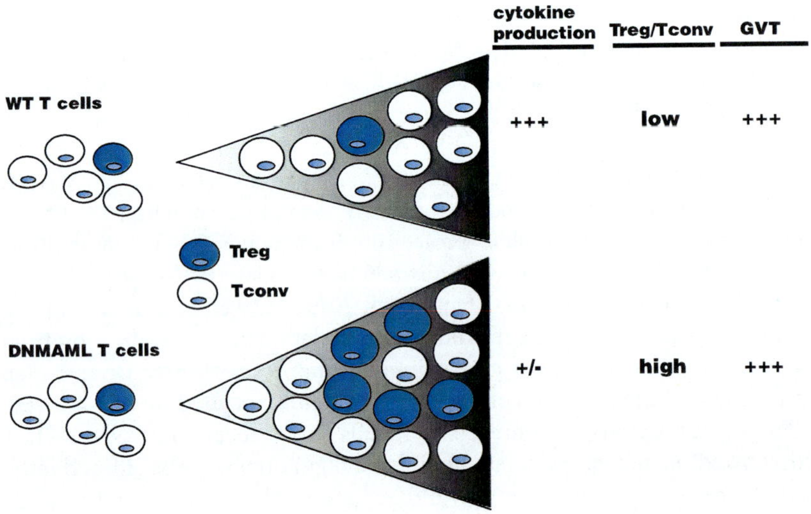

Fig. 3 Functional consequences of Notch blockade in alloreactive T cells during acute graft-versus-host disease. Notch inhibition prevents graft-versus-host disease without causing global immunosuppression (Zhang et al. 2011). DNMAML-mediated Notch blockade does not inhibit in vivo proliferation of alloreactive T cells. DNMAML T_{reg} cells expand at a greater rate than wild-type (WT) T_{reg} cells, leading to a higher T_{reg}/T_{conv} ratio in tissues of the transplant recipients. Alloreactive DNMAML T cells secrete markedly less inflammatory cytokines but maintain equivalent GVT activity when compared to alloreactive WT T cells

et al. 2011), or upon combined inactivation of *Notch1* and *Notch2* (unpublished data). Together, our genetic models indicate that Notch acts via canonical signaling through the CSL-RBP-Jκ/MAML transcriptional complex in alloreactive T cells. When compared to wild-type T cells, Notch signaling-deprived T cells produced reduced amounts of a broad range of inflammatory cytokines, including IFN-γ, IL-4, IL-17, TNFα, and IL-2 (Zhang et al. 2011) (Fig. 3). However, $CD4^+$ and $CD8^+$ DNMAML T-cells expressed normal levels of *Tbx21* (encoding T-bet) and elevated levels of *Eomes*, the master transcription factors regulating Th1 $CD4^+$ and effector $CD8^+$ T cell differentiation, respectively (Szabo et al. 2000; Pearce et al. 2003). These findings argue against a single T helper or cytotoxic T cell differentiation defect in Notch signaling-deprived alloreactive T cells, and also differ from findings in other systems (Maekawa et al. 2003; Minter et al. 2005; Maekawa et al. 2008; Okamoto et al. 2008; Amsen et al. 2004; Chao et al. 2009). We are actively investigating the novel mechanisms of Notch action and the transcriptional targets mediating the effects of the pathway in T cell alloimmunity.

The profound effects of Notch blockade in mouse models of GVHD suggest a strong potential for clinical translation. Short of an intervention to genetically modify donor T cells, Notch signaling modulation will require a pharmacologic approach. Gamma secretase inhibitors (GSIs) block a rate-limiting proteolytic step during Notch activation and have been developed to achieve systemic pan-Notch inhibition in animal models and in patients. However, on-target side effects have been problematic (van Es et al. 2005). After allo-BMT, the intestine is particularly sensitive due to its exposure to total body irradiation or other conditioning agents. Our observations in mouse models of GVHD indicate that GSIs lead to significant gut toxicity in this setting, without a good therapeutic window (unpublished data). As an alternative to GSIs, we are exploring the possibility of targeting individual Notch ligands and receptors in the immune system while sparing other organ systems that experience Notch signals mediated by other Notch pathway members.

In addition to its role in alloreactive T cells, recent work suggests that active Notch signaling within the DC compartment can also promote GVHD (Toubai et al. 2011). Bone marrow chimeric mice reconstituted with Ikaros-deficient bone marrow experienced an exacerbated form of GVHD when compared to chimeras reconstituted with control bone marrow. Ikaros-deficient DCs expressed higher levels of *Notch2*, *Notch3*, *Jagged1*, *Jagged2*, and Notch target genes. This was consistent with the activation of Notch signaling observed in other contexts upon Ikaros deficiency (Dumortier et al. 2006). Furthermore, allogeneic T cells cocultured with Ikaros-deficient DCs proliferated at a greater rate than when cocultured with wild-type DCs, an effect that was abrogated by GSI treatment. It remains to be investigated if Notch signaling can exert similar functions at physiological levels of signaling in Ikaros-sufficient DCs.

6 Conclusions and Future Directions

Altogether, work described in this review highlights the potent effects of Notch signaling on alloreactive T cell differentiation and function. Some of the conclusions reported to date appear contradictory. In particular, forced exposure to overexpressed Notch ligands has been described to induce tolerance (Wong et al. 2003; Yvon et al. 2003), while most of the in vivo results using Notch loss-of-function systems indicate a role for Notch in promoting rather than inhibiting alloreactivity (Riella et al. 2011; Zhang et al. 2011). This includes the remarkable in vivo activity of Notch blockade in multiple models of allo-BMT (Zhang et al. 2011). The reasons underlying these opposite outcomes are not clear. However, artificial gain-of-function systems are prone to generate findings that may not predict the nature of Notch activity in vivo in T cell alloimmunity and other T cell-dependent immune responses. In vivo loss-of-function systems targeted to defined immune cell types will be essential for fostering progress in the field. In addition, it will be important to investigate both the overall effects of Notch signaling and the contribution from individual receptors and ligands, which may not be mimicked adequately in vitro. We also believe that the spectrum and nature of Notch's effects in mature T cells are far from being fully unraveled. We predict that future work will identify essential new functions of the Notch signaling pathway with fundamental immunobiological importance, paving the way toward therapeutic interventions that target Notch signaling in T cell alloimmunity and other T cell-mediated disorders.

Acknowledgments Work on Notch signaling in the Maillard laboratory has been supported by a Damon Runyon Rachleff Innovation award, a Scholar award from the American Society of Hematology and funding from the National Institutes of Health (RO1 AI091627).

References

Alegre ML, Chong A (2009) Toll-like receptors (TLRs) in transplantation. Front Biosci (Elite Ed) 1:36–43

Amsen D, Antov A, Flavell RA (2009) The different faces of notch in T-helper-cell differentiation. Nat Rev Immunol 9:116–124

Amsen D, Antov A, Jankovic D, Sher A, Radtke F, Souabni A, Busslinger M, McCright B, Gridley T, Flavell RA (2007) Direct regulation of Gata3 expression determines the T helper differentiation potential of Notch. Immunity 27: 89−99

Amsen D, Blander JM, Lee GR, Tanigaki K, Honjo T, Flavell RA (2004) Instruction of distinct CD4 T helper cell fates by different notch ligands on antigen-presenting cells. Cell 117: 515-526

Anderson BE, McNiff JM, Jain D, Blazar BR, Shlomchik WD, Shlomchik MJ (2005) Distinct roles for donor- and host-derived antigen-presenting cells and costimulatory molecules in murine chronic graft-versus-host disease: requirements depend on target organ. Blood 105: 2227-2234

Auderset F, Schuster S, Coutaz M, Koch U, Desgranges F, Merck E, Macdonald HR, Radtke F, Tacchini-Cottier, F (2012) Redundant Notch1 and Notch2 Signaling Is Necessary for IFNgamma Secretion by T Helper 1 Cells During Infection with Leishmania major. PLoS Pathog 8, e1002560

Bugeon L, Gardner LM, Rose A, Gentle M, Dallman MJ (2008) Cutting edge: notch signaling induces a distinct cytokine profile in dendritic cells that supports T cell-mediated regulation and IL-2-dependent IL-17 production. J Immunol 181:8189–8193

Caton ML, Smith-Raska MR, Reizis B (2007) Notch-RBP-J signaling controls the homeostasis of CD8- dendritic cells in the spleen. J Exp Med 204:1653–1664

Cho OH, Shin HM, Miele L, Golde TE, Fauq A, Minter LM, Osborne BA (2009) Notch regulates cytolytic effector function in CD8+ T cells. J Immunol 182: 3380-3389

Ciofani M, Schmitt TM, Ciofani A, Michie AM, Cuburu N, Aublin A, Maryanski JL, Zuniga-Pflucker JC (2004) Obligatory role for cooperative signaling by pre-TCR and Notch during thymocyte differentiation. J Immunol 17:5230-5239

Ciofani M, Zuniga-Pflucker JC (2005) Notch promotes survival of pre-T cells at the beta-selection checkpoint by regulating cellular metabolism. Nat Immunol 6:881–888

Coghill JM, Carlson MJ, Moran TP, Serody JS (2008) The biology and therapeutic potential of natural regulatory T-cells in the bone marrow transplant setting. Leuk Lymphoma 49:18601869

Delaney C, Heimfeld S, Brashem-Stein C, Voorhies H, Manger RL, Bernstein ID (2010) Notch-mediated expansion of human cord blood progenitor cells capable of rapid myeloid reconstitution. Nat Med 16:232–236

Dumortier A, Jeannet R, Kirstetter P, Kleinmann E, Sellars M, dos Santos NR, Thibault, C, Barths J, Ghysdael J, Punt JA et al (2006) Notch activation is an early and critical event during T-Cell leukemogenesis in Ikaros-deficient mice. Mol Cell Biol 26, 209-220

Ellisen LW Bird J, West DC, Soreng AL, Reynolds TC, Smith SD, Sklar J (1991) TAN-1, the human homolog of the *Drosophila* notch gene, is broken by chromosomal translocations in T lymphoblastic neoplasms. Cell 66:649–661

Falkenburg JH, Warren EH (2011) Graft versus leukemia reactivity after allogeneic stem cell transplantation. Biol Blood Marrow Transplant 17:S33–38

Ferrara JL, Levine JE, Reddy P, Holler E (2009) Graft-versus-host disease. Lancet 373:1550–1561

Fiorini E Merck E, Wilson A, Ferrero I, Jiang W, Koch U, Auderset F, Laurenti E, Tacchini-Cottier F, Pierres M et al (2009) Dynamic regulation of notch 1 and notch 2 surface expression during T cell development and activation revealed by novel monoclonal antibodies. J Immunol 183:7212–7222

Fowlkes BJ, Robey EA (2002) A reassessment of the effect of activated notch1 on CD4 and CD8 T cell development. J Immunol 169:1817–1821

Fu T, Zhang P, Feng L, Ji G, Wang XH, Zheng MH, Qin HY, Chen DL, Wang WZ, Han H (2011) Accelerated acute allograft rejection accompanied by enhanced T-cell proliferation and attenuated Treg function in RBP-J deficient mice. Mol Immunol 48:751–759

Gao J, Graves S, Koch U, Liu S, Jankovic V, Buonamici S, El Andaloussi A, Nimer SD, Kee BL, Taichman R et al (2009) Hedgehog signaling is dispensable for adult hematopoietic stem cell function. Cell Stem Cell 4:548–558

Han H, Tanigaki K, Yamamoto N, Kuroda K, Yoshimoto M, Nakahata T, Ikuta K, Honjo T (2002) Inducible gene knockout of transcription factor recombination signal binding protein-J reveals its essential role in T versus B lineage decision. Int Immunol 14:637–645

Ho VT, Soiffer RJ (2001) The history and future of T-cell depletion as graft-versus-host disease prophylaxis for allogeneic hematopoietic stem cell transplantation. Blood 98:3192–3204

Hoyne GF,Le Roux I, Corsin-Jimenez M, Tan K, Dunne J, Forsyth LM, Dallman MJ, Owen MJ, Ish-Horowicz D, Lamb JR (2000) Serrate1-induced notch signalling regulates the decision between immunity and tolerance made by peripheral CD4(+) T cells. Int Immunol 12:177–185

Hozumi K, Negishi N, Suzuki D, Abe N, Sotomaru Y, Tamaoki N, Mailhos C, Ish-Horowicz D, Habu S, Owen MJ(2004) Delta-like 1 is necessary for the generation of marginal zone B cells but not T cells in vivo. Nat Immunol 5:638–644

Ito T, Schaller M, Hogaboam CM, Standiford TJ, Sandor M, Lukacs NW, Chensue SW, Kunkel SL (2009) TLR9 regulates the mycobacteria-elicited pulmonary granulomatous immune response in mice through DC-derived notch ligand delta-like 4. J Clin Invest 119:33–46

Jones SC, Murphy GF, Friedman TM, Korngold R (2003) Importance of minor histocompatibility antigen expression by nonhematopoietic tissues in a CD4+ T cell-mediated graft-versus-host disease model. J Clin Invest 112:1880–1886

Kawai T, Cosimi AB, Spitzer TR, Tolkoff-Rubin N, Suthanthiran M, Saidman SL, Shaffer J, Preffer FI, Ding R, Sharma V et al (2008) HLA-mismatched renal transplantation without maintenance immunosuppression. N Engl J Med 358:353–361
Koch U, Fiorini E, Benedito R, Besseyrias V, Schuster-Gossler K, Pierres M, Manley NR, Duarte A, Macdonald HR, Radtke F (2008) Delta-like 4 is the essential, nonredundant ligand for notch1 during thymic T cell lineage commitment. J Exp Med 205:2515–2523
Kolb HJ (2008) Graft-versus-leukemia effects of transplantation and donor lymphocytes. Blood 112:4371–4383
Kopan R, Ilagan MX (2009) The canonical notch signaling pathway: unfolding the activation mechanism. Cell 137:216–233
Koyama M, Kuns RD, Olver SD, Raffelt NC, Wilson YA, Don AL, Lineburg KE, Cheong M, Robb RJ, Markey KA et al (2012) Recipient nonhematopoietic antigen-presenting cells are sufficient to induce lethal acute graft-versus-host disease. Nat Med 18:135–142
Leventhal J, Abecassis M, Miller J, Gallon L, Ravindra K, Tollerud DJ, King B, Elliott MJ, Herzig G, Herzig R et al (2012) Chimerism and tolerance without GVHD or engraftment syndrome in HLA-mismatched combined kidney and hematopoietic stem cell transplantation. Sci Transl Med 4(124):ra28
Lewis KL, Caton ML, Bogunovic M, Greter M, Grajkowska LT, Ng D, Klinakis A, Charo IF, Jung S, Gommerman JL et al (2011) Notch2 receptor signaling controls functional differentiation of dendritic cells in the spleen and intestine. Immunity 35:780–791
Li H, Demetris AJ, McNiff J, Matte-Martone C, Tan HS, Rothstein DM, Lakkis FG, Shlomchik WD (2012) Profound Depletion of Host Conventional Dendritic Cells, Plasmacytoid Dendritic Cells, and B Cells Does Not Prevent Graft-versus-Host Disease Induction. J Immunol 188:3804–3811
Li XC, Rothstein DM, Sayegh MH (2009) Costimulatory pathways in transplantation: challenges and new developments. Immunol Rev 229:271–293
Louvi A, Artavanis-Tsakonas S (2012) Notch and disease: A growing field. Semin Cell Dev Biol (epub ahead of print)
Maekawa Y, Minato Y, Ishifune C, Kurihara T, Kitamura A, Kojima H, Yagita H, Sakata-Yanagimoto M, Saito T, Taniuchi I et al (2008) Notch2 integrates signaling by the transcription factors RBP-J and CREB1 to promote T cell cytotoxicity. Nat Immunol 9:1140–1147
Maekawa Y, Tsukumo S, Chiba S, Hirai H, Hayashi Y, Okada H, Kishihara K, Yasutomo K(2003) Delta1-notch3 interactions bias the functional differentiation of activated CD4+ T cells. Immunity 19:549–559
Maillard I, Fang T, Pear WS (2005) Regulation of lymphoid development, differentiation and function by the notch pathway. Annu Rev Immunol 23:945–974
Maillard I, Koch U, Dumortier A, Shestova O, Xu L, Sai H, Pross SE, Aster JC, Bhandoola A, Radtke F et al (2008) Canonical notch signaling is dispensable for the maintenance of adult hematopoietic stem cells. Cell Stem Cell 2:356–366
Maillard I,Tu L, Sambandam A, Yashiro-Ohtani Y, Millholland J, Keeshan K, Shestova O, Xu L, handoola A, Pear WS (2006) The requirement for notch signaling at the beta-selection checkpoint in vivo is absolute and independent of the pre-T cell receptor. J Exp Med 203:2239–2245
Maillard I, Weng AP, Carpenter AC, Rodriguez CG, Sai H, Xu L, Allman D, Aster JC, Pear WS (2004) Mastermind critically regulates notch-mediated lymphoid cell fate decisions. Blood 104:1696–1702
Minter LM, Turley DM, Das P, Shin HM, Joshi I, Lawlor RG, Cho OH, Palaga T, Gottipati S, Telfer JC et al (2005) Inhibitors of gamma-secretase block in vivo and in vitro T helper type 1 polarization by preventing notch upregulation of Tbx21. Nat Immunol 6:680–688
Napolitani G, Rinaldi A, Bertoni F, Sallusto F, Lanzavecchia A (2005) Selected toll-like receptor agonist combinations synergistically trigger a T helper type 1-polarizing program in dendritic cells. Nat Immunol 6:769–776
Nguyen VH, Zeiser R, Dasilva DL, Chang DS, Beilhack A, Contag CH, Negrin RS (2007) In vivo dynamics of regulatory T-cell trafficking and survival predict effective strategies to control graft-versus-host disease following allogeneic transplantation. Blood 109:2649–2656

Nguyen VH, Zeiser R, Negrin RS (2006) Role of naturally arising regulatory T cells in hematopoietic cell transplantation. Biol Blood Marrow Transplant 12:995–1009
Okamoto M, Takeda K, Joetham A, Ohnishi H, Matsuda H, Swasey CH, Swanson BJ, Yasutomo K, Dakhama A, Gelfand EW (2008) Essential role of notch signaling in effector memory CD8+ T cell-mediated airway hyperresponsiveness and inflammation. J Exp Med 205:1087–1097
Osborne BA, Minter LM (2007) Notch signalling during peripheral T-cell activation and differentiation. Nat Rev Immunol 7:64–75
Pearce EL, Mullen AC, Martins GA, Krawczyk CM, Hutchins AS, Zediak VP, Banica M, DiCioccio CB, Gross DA, Mao CA et al (2003) Control of effector CD8+ T cell function by the transcription factor Eomesodermin. Science 302:1041–1043
Radtke F, Fasnacht N, Macdonald HR (2010) Notch signaling in the immune system. Immunity 32:14–27
Radtke F, Wilson A, Stark G, Baue M, van Meerwijk J, MacDonald HR, Aguet M (1999) Deficient T cell fate specification in mice with an induced inactivation of notch1. Immunity 10:547–558
Riella LV, Ueno T, Batal I, De Serres SA, Bassil R, Elyaman W, Yagita H, Medina-Pestana JO, Chandraker A, Najafian N (2011) Blockade of notch ligand delta1 promotes allograft survival by inhibiting alloreactive Th1 cells and cytotoxic T cell generation. J Immunol 187:4629–4638
Robey E, Chang D, Itano A, Cado D, Alexander H, Lans D, Weinmaster G, Salmon P(1996) An activated form of notch influences the choice between CD4 and CD8 T cell lineages. Cell 87:483–492
Sambandam A Maillard I, Zediak VP, Xu L, Gerstein R, Aster J, Pear WS, Bhandoola A (2005) Notch signaling controls the generation and differentiation of early T lineage progenitors. Nat Immunol 6:663–670
Sandy AR, Maillard I (2009) Notch signaling in the hematopoietic system. Expert Opin Biol Ther 9:1383–1398
Schaller MA, Neupane R, Rudd BD, Kunkel SL, Kallal LE, Lincoln P, Lowe JB, Man Y, Lukacs NW (2007) Notch ligand delta-like 4 regulates disease pathogenesis during respiratory viral infections by modulating Th2 cytokines. J Exp Med 204:2925–2934
Shlomchik WD (2007) Graft-versus-host disease. Nat Rev Immunol 7:340–352
Shlomchik WD, Couzens MS, Tang CB, McNiff J, Robert ME, Liu J, Shlomchik MJ, Emerson SG (1999) Prevention of graft versus host disease by inactivation of host antigen-presenting cells. Science 285:412–415
Suchin EJ, Langmuir PB, Palmer E, Sayegh MH, Wells AD, Turka LA (2001) Quantifying the frequency of alloreactive T cells in vivo: new answers to an old question. J Immunol 166:973–981
Szabo SJ, Kim ST, Costa GL, Zhang X, Fathman CG, Glimcher LH (2000) A novel transcription factor, T-bet, directs Th1 lineage commitment. Cell 100:655–669
Tan JB, Visan I, Yuan JS, Guidos CJ (2005) Requirement for notch1 signals at sequential early stages of intrathymic T cell development. Nat Immunol 6:671–679
Tanigaki K, Tsuji M, Yamamoto N, Han H, Tsukada J, Inoue H, Kubo M, Honjo T (2004) Regulation of alphabeta/gammadelta T cell lineage commitment and peripheral T cell responses by notch/RBP-J signaling. Immunity 20:611–622
Tawara I, Shlomchik WD, Jones A, Zou W, Nieves E, Liu C, Toubai T, Duran-Struuck R, Sun Y, Clouthier SG et al (2010) A crucial role for host APCs in the induction of donor CD4+CD25+ regulatory T cell-mediated suppression of experimental graft-versus-host disease. J Immunol 185:3866–3872
Toubai T, Sun Y, Tawara I, Friedman A, Liu C, Evers R, Nieves E, Malter C, Chockley P, Maillard I et al (2011) Ikaros-notch axis in host hematopoietic cells regulates experimental graft-versus-host disease. Blood 118:192–204
Toubai T, Tawara I, Sun Y, Liu C, Nieves E, Evers R, Friedman T, Korngold R, Reddy P (2012) Induction of acute GVHD by sex-mismatched H-Y antigens in the absence of functional radiosensitive host hematopoietic-derived antigen-presenting cells. Blood 119:3844–3853
Tu L, Fang TC, Artis D, Shestova O, Pross SE, Maillard I, Pear WS (2005) Notch signaling is an important regulator of type 2 immunity. J Exp Med 202:1037–1042

van Es JH, van Gijn ME, Riccio O, van den Born M, Vooijs M, Begthel H, Cozijnsen M, Robine S, Winton DJ, Radtke F et al (2005) Notch/gamma-secretase inhibition turns proliferative cells in intestinal crypts and adenomas into goblet cells. Nature 435:959–963

Varnum-Finney B, Brashem-Stein C, Bernstein ID (2003) Combined effects of notch signaling and cytokines induce a multiple log increase in precursors with lymphoid and myeloid reconstituting ability. Blood 101:1784–1789

Varnum-Finney B Xu L, Brashem-Stein C, Nourigat C, Flowers D, Bakkour S, Pear WS, Bernstein ID (2000) Pluripotent, cytokine-dependent, hematopoietic stem cells are immortalized by constitutive notch1 signaling. Nat Med 6:1278–1281

Vigouroux S ,Yvon E, Wagner HJ, Biagi E, Dotti G, Sili U, Lira C, Rooney CM, Brenner MK (2003) Induction of antigen-specific regulatory T cells following overexpression of a notch ligand by human B lymphocytes. J Virol 77:10872–10880

Welniak LA, Blazar BR, Murphy WJ (2007) Immunobiology of allogeneic hematopoietic stem cell transplantation. Annu Rev Immunol 25:139–170

Weng AP, Ferrando AA, Lee W, Morris JPt, Silverman LB, Sanchez-Irizarry C, Blacklow SC, Look AT, Aster JC (2004) Activating mutations of NOTCH1 in human T cell acute lymphoblastic leukemia. Science 306:269–271

Weng AP, Nam Y, Wolfe MS, Pear WS, Griffin JD, Blacklow SC, Aster JC (2003) Growth suppression of pre-T acute lymphoblastic leukemia cells by inhibition of notch signaling. Mol Cell Biol 23:655–664

Wolfer A, Bakker T, Wilson A, Nicolas M, Ioannidis V, Littman DR, Wilson CB, Held W, MacDonald HR, Radtke F (2001) Inactivation of notch 1 in immature thymocytes does not perturb CD4 or CD8T cell development. Nat Immunol 2:235–241

Wong KK, Carpenter MJ, Young LL, Walker SJ, McKenzie G, Rust AJ, Ward G, Packwood L, Wahl K, Delriviere L et al (2003) Notch ligation by delta1 inhibits peripheral immune responses to transplantation antigens by a CD8+ cell-dependent mechanism. J Clin Invest 112:1741–1750

Wu CJ, Ritz J (2006) Induction of tumor immunity following allogeneic stem cell transplantation. Adv Immunol 90:133–173

Wysocki CA, Panoskaltsis-Mortari A, Blazar BR, Serody JS (2005) Leukocyte migration and graft-versus-host disease. Blood 105:4191–4199

Yashiro-Ohtani Y, He Y, Ohtani T, Jones ME, Shestova O, Xu L, Fang TC, Chiang MY, Intlekofer AM, Blacklow SC et al (2009) Pre-TCR signaling inactivates notch1 transcription by antagonizing E2A. Genes and Development 23:1665–1676

Yuan JS, Kousis PC, Suliman S, Visan I, Guidos CJ (2010) Functions of notch signaling in the immune system: consensus and controversies. Annu Rev Immunol 28:343–365

Yvon ES, Vigouroux S, Rousseau RF, Biagi E, Amrolia P, Dotti G, Wagner HJ, Brenner MK (2003) Over expression of the notch ligand, jagged-1 induces alloantigen-specific human regulatory T cells. Blood 102:3815–3821

Zhang Y, Louboutin JP, Zhu J, Rivera AJ, Emerson SG (2002) Preterminal host dendritic cells in irradiated mice prime CD8+ T cell-mediated acute graft-versus-host disease. J Clin Invest 109:1335–1344

Zhang Y, Sandy AR, Wang J, Radojcic V, Shan GT, Tran IT, Friedman A, Kato K, He S, Cui S et al (2011) Notch signaling is a critical regulator of allogeneic CD4+ T-cell responses mediating graft-versus-host disease. Blood 117:299–308

Notch2 and Immune Function

Mamiko Sakata-Yanagimoto and Shigeru Chiba

Abstract Notch2 is expressed in many cell types of most lineages in the hematolymphoid compartment and has specific roles in differentiation and function of various immune cells. Notch2 is required for development of splenic marginal zone B cells and regulates differentiation of dendritic cells (DCs) in the spleen. Notch2 appears to play some specific roles in the intestinal immunity, given that the fate of mast cells and a subset of DCs is regulated by Notch2 in the intestine. Notch2 also has important roles in helper T cell divergence from na CD4 T cells and activation of cytotoxic T cells. Moreover, recent genetic evidence suggests that both gain-and loss-of-function abnormalities of Notch2 cause transformation of immune cells. Inactivating mutations are found in Notch2 signaling pathways in chronic myelomonocytic leukemia, while activating mutations are found in mature B cell lymphomas, which reflects the role of Notch2 in the developmental process of these cells.

Contents

M. Sakata-Yanagimoto · S. Chiba (✉)
Department of Hematology, Faculty of Medicine, University of Tsukuba,
1-1-1 Tennodai, Tsukuba, Ibaraki 305-8575, Japan
e-mail: schiba-tky@umin.net

Current Topics in Microbiology and Immunology (2012) 360: 151–161
DOI: 10.1007/82_2012_235

Published Online: 14 June 2012

1 Introduction

The expression pattern of Notch1, Notch2, Notch3, and Notch4 varies from one cell type to another. Notch2 plays specific roles in the immune compartment independently of and cooperatively with Notch1 and Notch3. In this chapter, we mostly focus on evidence based on mouse genetic studies regarding Notch2 functions in immune cells. In the last part, we discuss the involvement of Notch2 in neoplastic transformation in conjunction with human hematologicmalignancies.

2 Notch2 Signaling in Marginal Zone B Cell Development

Mature splenic B cells are mainly divided into 2 types of B cells, follicular B (FOB) cells and marginal zone B (MZBs) cells (Martin and Kearney 2002). FOB cells are one of the main effectors of acquired immunity, able to respond to a large variety of antigens, while MZB cells can only elicit an immune response to a limited number of antigens. Immature B cells, developing from hematopoietic stem cells (HSCs) in the bone marrow, migrate to the spleen, and differentiate first into T1 (type1) transitional B cells (characterized by: IgM^{hi} IgD^{lo} $CD21^{lo}$), before differentiating into T2 (type2) transitional B cells (IgM^{hi} IgD^{hi} $CD21^{int}$ $CD23^{hi}$) (Loder et al. 1999). These progenitors further differentiate into the two types of B cells, MZB cells (IgM^{hi} IgD^{lo} $CD21^{hi}$ $CD23^{lo}$) and FOB cells (IgM^{lo} IgD^{hi} $CD21^{int}$ $CD23^{hi}$) (Martin and Kearney 2002). Conditional inactivation of *Notch2* in the B cell lineage results in defective MZB cell development, while FOB cell development is unaffected (Saito et al. 2003), which is basically consistent with the phenotype of *RBP-J* conditional knockout mice (Tanigaki et al. 2002). The cleaved Notch-RBP-J activator complex contains at least one out of three family members of the mastermind-like proteins (MAML1-3). Mastermind-like1 (MAML1) plays an essential role in MZB cell development, which is why MZB cells are lacking in *MAML1*-null mice (Wu et al. 2007; Oyama et al. 2007). Among Notch ligands, Delta-like 1 (Dll1) is responsible for MZB cell development, based on the fact that *Dll1* deletion using the Mx-Cre *loxP* system leads to loss of MZB cells (Hozumi et al. 2004; Sheng et al. 2008). Several lines of evidence indicate that loss of *Dll1* expression on nonhematopoietic cells causes MZB cell defects (Hozumi et al. 2004; Sheng et al. 2008; Tan et al. 2009). However, the exact cell types through which Dll1-induced Notch2 signaling triggers MZB development remains to be elucidated. The essential role of Notch signaling in MZB cell development is further proven by a number of other gene-targeted mice in which

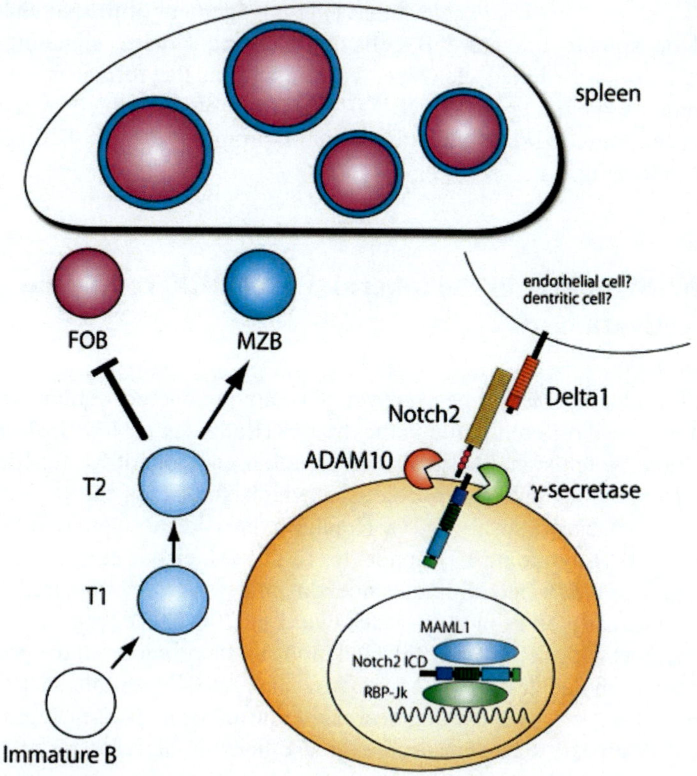

Fig. 1 Notch2 regulates marginal zone B cell development. Dll1 engagement of the Notch2 receptor expressed on splenic immature B cells initiates the Notch2 signaling cascade through ADAM10-mediated cleavage. This results in the formation of Notch-RBP-J-MAML1 proteins, which regulates gene expression and skews the differentiation program toward MZB cells rather than FOB cells

genes of the Notch signaling cascade were inactivated. Mind bomb (Mib), an E3 ubiquitin ligase, activates Notch signaling through endocytosis of Notch ligands. Deletion of *Mib1* in nonhematopoietic cells recapitulates defects in MZBs, whereas deletion in hematopoietic cells resulted in MZB levels that were comparable to control mice (Song et al. 2008).

Deletion of *ADAM10*, which encodes a matrix metalloprotease that processes the extracellular domain of the Notch receptor after ligand binding, also results in defects in MZB cells (Gibb et al. 2010). In contrast, deletion of *Msx2-interacting nuclear target protein* (*MINT*), which is a repressor of RBP-J mediated transcriptional activity and thus considered to be a negative regulator for Notch signaling, showed decreased numbers of FOB cells and increased numbers of MZB cells (Kuroda et al. 2003). Notch receptors can be modified by fringe glycosyl transferases. Lunatic and manic fringe were shown to cooperatively enhance Dll1-Notch2 interaction, and thereby induce MZB development (Tan et al. 2009).

Taken together, these data suggest that Dll1 engagement of the Notch2 receptor expressed on splenic immature B cells initiates the Notch2 signaling cascade through ADAM10-mediated cleavage. This results in the formation of a multi-protein complex including Notch-RBP-J-MAML1 proteins, which regulates gene expression and thereby skews the differentiation program toward MZB cells rather than FOB cells (Fig. 1).

3 Notch2 Signaling in Peripheral T Cell Differentiation and Activation

Signaling through Notch1 has been proven to be among the most important systems for immature T cell differentiation in the thymus (Radtke et al. 1999). During T cell differentiation, Notch signaling is also an essential component for functional maturation and activation of peripheral T cells, in which Notch2 appears to be among the main players (Maekawa et al. 2008). Notch2 expression is increased along with activation of CD8+ cytotoxic T lymphocytes (CTL) (Maekawa et al. 2008). CTLs of both *Notch2* and *RBP-J* conditional knockout mice show an impaired activation potential in vitro as well as in vivo (Maekawa et al. 2008). Cleaved Notch2 (N2IC) directly interacts with CREB and p300 and binds to the promoter of the granzyme B gene, an effector molecule of CTL (Maekawa et al. 2008). Conditional inactivation of *Notch2* in CD8+ T cells results in a decreased antitumor response (Sugimoto et al. 2010). Notch1 appears to be dispensable for an efficient CTL response given the fact that deletion of *Notch1* in CD8+ T cells shows an antitumor response comparable to control mice (Sugimoto et al. 2010). However, this view was recently challenged by a report showing that Notch1 also directly controls main players of CTL, including Eomes, perforin, and granzyme B (Cho et al. 2009).

Notch2/Notch1 double deficient animals reveal impaired differentiation of na CD4+ T cells toward helper T2 (Th2) cells. Notch was shown to directly regulate the transcription of the transcription factor GATA3, and the cytokine interleukin-4 (IL-4) (Amsen et al. 2004, 2007; Fang et al. 2007), both of which are important mediators of Th2 differentiation. *RBP-J* deficient animals recapitulate the phenotype observed in *Notch2/Notch1* deficient animals indicating that this process is mediated through canonical Notch signaling (Amsen et al. 2004). The role of Notch signaling in Th1 differentiation is less clear. Several reports demonstrated that Th1 differentiation is augmented by Notch signaling (Maekawa et al. 2003; Sun et al. 2008); however, a Th1 response is maintained in *Notch2/Notch1* double deficient, and *RBP-J* deficient animals (Amsen et al. 2004), as well as in mice expressing a dominant negative *MAML1* (Tu et al. 2005), questioning the importance of Notch signaling in Th1 differentiation.

With certainty, it can thus be summarized that Notch2 signaling induces cytotoxic T cell differentiation and activation, and that Notch1 and Notch2 concordantly induce Th2 cell differentiation.

4 Notch2 Signaling in Mast Cells

Mast cells arise from HSCs in the bone marrow, migrate to peripheral tissues as immature progenitors, where they subsequently differentiate into mature mast cells (Galli et al. 2005). However, the detailed process of their development is still disputed. Mast cells can be generated in vitro by culturing mouse bone marrow cells with a defined cocktail of cytokines. This in vitro system allows to partially mimic the physiologic development of mast cells. Notch2 signaling instructs myeloid progenitors to adopt a mast cell fate as opposed to differentiate into neutrophils or macrophages, through the coordinated regulation of *Hes1* and *GATA3* (Sakata-Yanagimoto et al. 2008). Mast cells are divided into two subtypes; mucosal and connective tissue type mast cells (Gurish and Boyce 2006; Miller and Pemberton 2002). Each subtype features specific mast-cell proteases (mMCP) (Miller and Pemberton 2002). Notch2 signaling skews cultured mast cells toward the mucosal type rather than connective tissue type (M.S.-Y. and S.C., unpublished data). The *Strongyloides venezuelensis* (*SV*) infection model is useful for analyzing mast cell-mediated mucosal immunity (Maruyama et al. 2000). This nematode evokes intraepithelial mast cell hyperplasia in the small intestine (Maruyama et al. 2000). *Notch2*-null mice show impaired expulsion of *SV*, possibly because of a delayed mast cell progenitor production in the bone marrow, impaired migration of mast cells from the lamina propria to the intraepithelium of the intestine, and impaired activation of intestinal mast cells (Sakata-Yanagimoto et al. 2011). The number and distribution of connective tissue-type mast cells are normal in *Notch2*-null mice (Sakata-Yanagimoto et al. 2011), suggesting that Notch2 signaling is specifically required for proper migration and activation of intestinal mast cells.

5 Notch2 Signaling in Dendritic Cells

Dentritic cells (DCs) initiate immune responses by presenting antigen to na T cells (Steinman and Idoyaga 2010). DCs arise from common bone marrow progenitors that can give rise to both DCs and macrophages (Steinman and Idoyaga 2010). DCs comprise two subclasses, i.e., the so-called plasmacytoid DCs and classical DCs. Classical DCs residing in the spleen are further classified into two main subsets; CD8+CD11b− DCs which mediate cross-presentation to cytotoxic T cells via MHC class I pathway (Dudziak et al. 2007; den Haan et al. 2000) and CD8−CD11b+ DCs which preferentially present MHC class II restricted antigens to CD4+ helper T cells (Dudziak et al. 2007). CD8−CD11b+ DCs are mainly localized in the marginal zone, adjacent to the Dll1-expressing cells (Caton et al. 2007). CD11b+ DCs in the lamina propria of the intestine contain two distinct subsets; CD11b+CD103+ DCs and CD11b+CD103− DCs. CD11b+CD103+ DCs migrate to mesenteric lymph nodes and are presumed to have antigen presenting potential to helper T cells (Denning et al. 2011; Bogunovic et al. 2009).

DC-specific deletion of either *Notch2* or *RBP-J* impairs the development of CD8−CD11b+ DCs in the spleen (Caton et al. 2007; Lewis et al. 2011). CD8+ DCs are also decreased by deletion of *Notch2* but are not affected by the deletion of *RBP-J* (Caton et al. 2007; Lewis et al. 2011). Splenic CD11b+ DCs are divided into two subsets according to the expression levels of Esam and Cx3cr1 (Lewis et al. 2011). CD11b+Esam^{high}Cx3cr1^{low} but not CD11b+Esam^{low}Cx3cr1^{high} DCs are almost abrogated by deletion of either *Notch2* or *RBP-J* (Lewis et al. 2011). CD11b+Esam^{high}Cx3cr1^{low} cells are required for proper priming of T cells in the spleen, which are reduced in *RBP-J*-null mice (Lewis et al. 2011).

Notch2 selectively controls CD11b+CD103+ DCs in the lamina propria of the intestine as well as those that migrate toward mesenteric lymph nodes, which in turn are important for supporting IL-17 producing CD4+ T cells. CD11b+CD103+ DCs are not affected by the inactivation of *RBP-J* (Lewis et al. 2011).

Taken together, Notch2 regulates tissue-specific subsets of DCs in the spleen and in the intestine. Notch2 function might be partly mediated by a RBP-J independent/noncannonical pathway.

6 Notch2 Signaling in Hematopoietic Stem Cells

Notch signaling plays an essential role in self-renewal of stem cells as well as in the growth and differentiation of diverse progenitors within various organs. In contrast, the role of Notch signaling in self-renewal of HSC has been disputed over the years. Early in vitro gain-of-function experiments, such as introduction of a constitutive active form of Notch (Varnum-Finney et al. 2000; Stier et al. 2002) or the transcription factor Hes1 (Kunisato et al. 2003), and stimulation of HSCs with cell-surface expressed ligands or ligand-immunoglobulin chimeric proteins (Karanu et al. JEM 2000; Ohisi et al. JCI 2002; Suluki et al. Stem cells 2006), indicated that Notch signaling supports self-renewal of HSCs and has a role in HSC expansion. On the contrary, several loss-of-function experiments suggest that Notch signaling is dispensable for maintenance of HSCs. HSCs lacking *RBP-J* and those expressing dominant negative *MAML1*, a potent inhibitor of the Notch transcriptional complex, achieve long-term reconstitution comparable to wild-type HSCs, when transplanted into irradiated mice (Maillard et al. 2008). The reconstitution potential of HSCs null for both *Notch1* and *Jagged1* was shown to be comparable to that of wild-type HSCs (Mancini et al. 2005).

However, recently, such negative findings were partially challenged. At a very early time point after treatment with 5-fluorouracil, the number of multipotent progenitors (MPPs) was decreased in *Notch2*-null mice, compared to that in control mice (Varnum-Finney et al. 2011). Similarly, shortly after transplantation, both MPPs and long-term HSCs were decreased in *Notch2*-null BM transplanted mice (Varnum-Finney et al. 2011).

In summary, these data suggest that although Notch signaling is dispensable for homeostasis of HSCs, in challenge and stress situations signaling through Notch2 seems to play a role in the process of HSCs expansion.

7 Notch2 Signaling in Transformation of Blood Cells

7.1 Notch2 Mutations in B Cell Lymphomas

Notch1 is among the most important molecules for physiologic development of T cells (Radtke et al. 1999), and Notch2 is indispensable for MZB cell development (Saito et al. 2003) as described above. Discovery of hyperactivation of Notch1 and Notch2 through gain-of-function mutations in immature T cell neoplasms (Weng et al. 2004) (T cell acute lymphoblastic leukemia or T-ALL in humans) and in subtypes of mature B cell neoplasms (Lee et al. 2009; Troen et al. 2008), respectively, appears to echo the physiologic roles of these molecules in specific lineages and differentiation stages. Those mutations are concentrated in the extracellular heterodimerization (HD) domain and the intracellular proline-, glutamic acid-, serine-, and threonine-rich (PEST) domain of *Notch1* in T-ALL (Weng et al. 2004), and only in the PEST domain of *Notch2* in mature B cell lymphomas (Lee et al. 2009; Troen et al. 2008). The distribution of mutations suggests that hyperactivation of Notch2 signaling in B cell lymphomas still requires binding of the ligand, whereas mutations within the HD domain of *Notch1* in T-ALL results in ligand independent activation of Notch1 signaling. In contrast to the fact that *Notch1* mutations are found in approximately 50 % of T-ALL cases (Weng et al. 2004), *Notch2* mutations were identified in only five out of 63 cases (8 %) of diffuse B-cell lymphoma (Lee et al. 2009) and in two out of 41 cases (5 %) of MZB cell lymphoma (Troen et al. 2008).

The relationship between B cell development and gain-of-function mutations in Notch2 is not as clear as in the context of T cell development and Notch1 mutations. Genetic evidence described above strongly suggests an oncogenic role of deregulated Notch2 in B lineage transformations. On the contrary, there has been a series of reports describing the tumor suppressive function of Notch signaling in B lineage cells, particularly in B-cell lymphoblastic leukemia (B-ALL) (Zweidler-McKay et al. 2005; Kannan et al. 2011), although loss-of-function mutations have not been found in the Notch2 signaling pathway. Integrating these pieces of information, it seems likely that Notch2 signaling can context dependently promote or suppress growth of B lineage cells. Another complexity was recently added by the identification of Notch1 mutations through the genome-wide screening of patient samples suffering from chronic lymphocytic leukemia (Puente et al. 2011), a type of intermediately mature B cell neoplasm and mantle cell lymphoma (Kridel et al. 2012), another type of mature B neoplasms.

7.2 Notch2 Signaling in Myeloid Neoplasms

Recently, Notch signaling was proven to function as a tumor-suppressor in chronic myelomonocytic leukemia (CMML) (Klinakis et al. 2011); several components of the Notch pathway, including *Nicastrin* (*NCSTN*), *APH1A*, *MAML1*, and *Notch2*

Table 1 Role of Notch signaling as tumor activator or tumor suppressor in hematopoietic leukemia/lymphoma

	Gain-of-function			Loss-of-function	
	T	B	M	B	M
Notch1	T-ALL ATL	CLL MCL DLBCL	AML	(B-ALL)[a]	
Notch2		MZB lymphoma DLBCL			CMML

T-ALL, T-cell acute lymphocytic leukemia; ATL, adult T-cell leukemia/lymphoma; CLL, chronic lymphocytic leukemia; MCL, mantle cell lymphoma; DLBCL, diffuse large B-cell lymphoma; MZB lymphoma, marginal zone B-cell lymphoma; AML, acute myeloid leukemia; B-ALL, B-cell acute lymphocytic leukemia; CMML, chronic myelomonocytic leukemia

[a] Loss-of-function mutations in Notch signal components have not been found in B-ALL

itself were found to be mutated and defective in CMML patients. This conclusion is also supported by the phenotype of mice lacking *NCSTN*, a component of γ-secretase, as well as that of *Notch1*-, *Notch2*-, and *Notch3*- triple null mice (Klinakis et al. 2011). These animals show enhanced granulocyte-monocyte progenitor potential and develop a fatal CMML-like disease (Klinakis et al. 2011). On the contrary, activating mutations of Notch1 were found in acute myeloid leukemia, a precursor myeloid neoplasms, though the frequency is less than 1 % (Wouters et al. 2007; Fu et al. 2006). Thus, as is the case of B cell malignancies, Notch signaling can function as either tumor promoter or suppressor within myeloid neoplasms.

These oncogenic and tumor suppressive functions of Notch1 and Notch2 signaling in T cell, B cell, and myeloid lineages have been summarized in Table 1. Knowledge about this area will expand rapidly in the very near future using current sequencing technology.

8 Conclusion

Signaling through Notch2 has an essential role in two major cell types present in the marginal zone of the spleen, splenic MZB cells, and splenic DCs. Notch2 signaling also mediates intestinal immunity by regulating development and localization of intestinal DCs and mast cells, and development of helper T cells and CTLs. Genetic and biologic evidence indicates that abnormal Notch2 signaling is involved in transformation of immune cells, although its functions appear to be bivalent; oncogenic signaling for mature B neoplasms and tumor suppressive signaling for mature myeloid neoplasms. The reason of specificity and non-redundant functions of Notch2 in the immune system may be partly attributed to the differences in expression patterns among Notch family genes, although this issue needs to be elucidated in more detail in future studies.

References

Amsen D, Blander JM, Lee GR, Tanigaki K, Honjo T, Flavell RA (2004) Instruction of distinct CD4 T helper cell fates by different notch ligands on antigen-presenting cells. Cell 117(4):515–526

Amsen D, Antov A, Jankovic D et al (2007) Direct regulation of Gata3 expression determines the T helper differentiation potential of Notch. Immunity 27(1):89–99

Bogunovic M, Ginhoux F, Helft J et al (2009) Origin of the lamina propria dendritic cell network. Immunity 31(3):513–525

Caton ML, Smith-Raska MR, Reizis B (2007) Notch-RBP-J signaling controls the homeostasis of CD8− dendritic cells in the spleen. J Exp Med 204(7):1653–1664

Cho OH, Shin HM, Miele L et al (2009) Notch regulates cytolytic effector function in CD8+ T cells. J Immunol 182(6):3380–3389

den Haan JM, Lehar SM, Bevan MJ (2000) CD8(+) but not CD8(−) dendritic cells cross-prime cytotoxic T cells in vivo. J Exp Med 192(12):1685–1696

Denning TL, Norris BA, Medina-Contreras O et al (2011) Functional specializations of intestinal dendritic cell and macrophage subsets that control Th17 and regulatory T cell responses are dependent on the T cell/APC ratio, source of mouse strain, and regional localization. J Immunol 187(2):733–747

Dudziak D, Kamphorst AO, Heidkamp GF et al (2007) Differential antigen processing by dendritic cell subsets in vivo. Science 315(5808):107–111

Fang TC, Yashiro-Ohtani Y, Del Bianco C, Knoblock DM, Blacklow SC, Pear WS (2007) Notch directly regulates Gata3 expression during T helper 2 cell differentiation. Immunity 27(1):100–110

Fu L, Kogoshi H, Nara N, Tohda S (2006) NOTCH1 mutations are rare in acute myeloid leukemia. Leuk Lymphoma 47(11):2400–2403

Galli SJ, Nakae S, Tsai M (2005) Mast cells in the development of adaptive immune responses. Nat Immunol 6(2):135–142

Gibb DR, El Shikh M, Kang DJ et al (2010) ADAM10 is essential for Notch2-dependent marginal zone B cell development and CD23 cleavage in vivo. J Exp Med 207(3):623–635

Gurish MF, Boyce JA (2006) Mast cells: ontogeny, homing, and recruitment of a unique innate effector cell. J Allergy Clin Immunol 117(6):1285–1291

Hozumi K, Negishi N, Suzuki D et al (2004) Delta-like 1 is necessary for the generation of marginal zone B cells but not T cells in vivo. Nat Immunol 5(6):638–644

Karanu FN et al. (2000) The notch ligand Jagged-1 represents a novel growth factor of human hematopoietic stem cells. JEM 192 (9) 1365–1372

Kannan S, Fang W, Song G et al (2011) Notch/HES1-mediated PARP1 activation: a cell type-specific mechanism for tumor suppression. Blood 117(10):2891–2900

Klinakis A, Lobry C, Abdel-Wahab O et al (2011) A novel tumour-suppressor function for the Notch pathway in myeloid leukaemia. Nature 473(7346):230–233

Kridel R, Meissner B, Rogic S et al (2012) Whole transcriptome sequencing reveals recurrent NOTCH1 mutations in mantle cell lymphoma. Blood 119(9):1963–1971

Kunisato A, Chiba S, Nakagami-Yamaguchi E et al (2003) HES-1 preserves purified hematopoietic stem cells ex vivo and accumulates side population cells in vivo. Blood 101(5):1777–1783

Kuroda K, Han H, Tani S et al (2003) Regulation of marginal zone B cell development by MINT, a suppressor of Notch/RBP-J signaling pathway. Immunity 18(2):301–312

Lee SY, Kumano K, Nakazaki K et al (2009) Gain-of-function mutations and copy number increases of Notch2 in diffuse large B-cell lymphoma. Cancer Sci 100(5):920–926

Lewis KL, Caton ML, Bogunovic M et al (2011) Notch2 receptor signaling controls functional differentiation of dendritic cells in the spleen and intestine. Immunity 35(5):780–791

Loder F, Mutschler B, Ray RJ et al (1999) B cell development in the spleen takes place in discrete steps and is determined by the quality of B cell receptor-derived signals. J Exp Med 190(1):75–89

Maekawa Y, Tsukumo S, Chiba S et al (2003) Delta1-Notch3 interactions bias the functional differentiation of activated CD4+ T cells. Immunity 19(4):549–559

Maekawa Y, Minato Y, Ishifune C et al (2008) Notch2 integrates signaling by the transcription factors RBP-J and CREB1 to promote T cell cytotoxicity. Nat Immunol 9(10):1140–1147

Maillard I, Koch U, Dumortier A et al (2008) Canonical notch signaling is dispensable for the maintenance of adult hematopoietic stem cells. Cell Stem Cell 2(4):356–366

Mancini SJ, Mantei N, Dumortier A, Suter U, MacDonald HR, Radtke F (2005) Jagged1-dependent Notch signaling is dispensable for hematopoietic stem cell self-renewal and differentiation. Blood 105(6):2340–2342

Martin F, Kearney JF (2002) Marginal-zone B cells. Nat Rev Immunol 2(5):323–335

Maruyama H, Yabu Y, Yoshida A, Nawa Y, Ohta N (2000) A role of mast cell glycosaminoglycans for the immunological expulsion of intestinal nematode, *Strongyloides venezuelensis*. J Immunol 164(7):3749–3754

Miller HR, Pemberton AD (2002) Tissue-specific expression of mast cell granule serine proteinases and their role in inflammation in the lung and gut. Immunology 105(4):375–390

Ohishi K et al. (2002) Delta-1 enhances marrow and thymus repopulating ability of human CD34(+)CD38(-) cord blood cells. J Clin Invest 110(8):1165–1174

Oyama T, Harigaya K, Muradil A et al (2007) Mastermind-1 is required for Notch signal-dependent steps in lymphocyte development in vivo. Proc Natl Acad Sci USA 104(23):9764–9769

Puente XS, Pinyol M, Quesada V et al (2011) Whole-genome sequencing identifies recurrent mutations in chronic lymphocytic leukaemia. Nature 475(7354):101–105

Radtke F, Wilson A, Stark G et al (1999) Deficient T cell fate specification in mice with an induced inactivation of Notch1. Immunity 10(5):547–558

Saito T, Chiba S, Ichikawa M et al (2003) Notch2 is preferentially expressed in mature B cells and indispensable for marginal zone B lineage development. Immunity 18(5):675–685

Sakata-Yanagimoto M, Nakagami-Yamaguchi E, Saito T et al (2008) Coordinated regulation of transcription factors through Notch2 is an important mediator of mast cell fate. Proc Natl Acad Sci USA 105(22):7839–7844

Sakata-Yanagimoto M, Sakai T, Miyake Y et al (2011) Notch2 signaling is required for proper mast cell distribution and mucosal immunity in the intestine. Blood 117(1):128–134

Sheng Y, Yahata T, Negishi N et al (2008) Expression of Delta-like 1 in the splenic non-hematopoietic cells is essential for marginal zone B cell development. Immunol Lett 121(1):33–37

Song R, Kim YW, Koo BK et al (2008) Mind bomb 1 in the lymphopoietic niches is essential for T and marginal zone B cell development. J Exp Med 205(11):2525–2536

Steinman RM, Idoyaga J (2010) Features of the dendritic cell lineage. Immunol Rev 234(1):5–17

Stier S, Cheng T, Dombkowski D, Carlesso N, Scadden DT (2002) Notch1 activation increases hematopoietic stem cell self-renewal in vivo and favors lymphoid over myeloid lineage outcome. Blood 99(7):2369–2378

Sugimoto K, Maekawa Y, Kitamura A et al (2010) Notch2 signaling is required for potent antitumor immunity in vivo. J Immunol 184(9):4673–4678

Sun J, Krawczyk CJ, Pearce EJ (2008) Suppression of Th2 cell development by Notch ligands Delta1 and Delta4. J Immunol 180(3):1655–1661

Suluki T et al. (2006) Highly efficient ex vivo expansion of human hematopoietic stem cells using Delta1-Fc chimeric protein. Stem Cells 24(11):2456–2465

Tan JB, Xu K, Cretegny K et al (2009) Lunatic and manic fringe cooperatively enhance marginal zone B cell precursor competition for delta-like 1 in splenic endothelial niches. Immunity 30(2):254–263

Tanigaki K, Han H, Yamamoto N et al (2002) Notch-RBP-J signaling is involved in cell fate determination of marginal zone B cells. Nat Immunol 3(5):443–450

Troen G, Wlodarska I, Warsame A, Hernandez Llodra S, De Wolf-Peeters C, Delabie J (2008) NOTCH2 mutations in marginal zone lymphoma. Haematologica 93(7):1107–1109

Tu L, Fang TC, Artis D et al (2005) Notch signaling is an important regulator of type 2 immunity. J Exp Med 202(8):1037–1042

Varnum-Finney B, Xu L, Brashem-Stein C et al (2000) Pluripotent, cytokine-dependent, hematopoietic stem cells are immortalized by constitutive Notch1 signaling. Nat Med 6(11):1278–1281

Varnum-Finney B, Halasz LM, Sun M, Gridley T, Radtke F, Bernstein ID (2011) Notch2 governs the rate of generation of mouse long- and short-term repopulating stem cells. J Clin Invest 121(3):1207–1216

Weng AP, Ferrando AA, Lee W et al (2004) Activating mutations of NOTCH1 in human T cell acute lymphoblastic leukemia. Science 306(5694):269–271

Wouters BJ, Jorda MA, Keeshan K et al (2007) Distinct gene expression profiles of acute myeloid/T-lymphoid leukemia with silenced CEBPA and mutations in NOTCH1. Blood 110(10):3706–3714

Wu L, Maillard I, Nakamura M, Pear WS, Griffin JD (2007) The transcriptional coactivator Maml1 is required for Notch2-mediated marginal zone B-cell development. Blood 110(10):3618–3623

Zweidler-McKay PA, He Y, Xu L et al (2005) Notch signaling is a potent inducer of growth arrest and apoptosis in a wide range of B-cell malignancies. Blood 106(12):3898–3906

Recent Advances on NOTCH Signaling in T-ALL

Gannie Tzoneva and Adolfo A. Ferrando

Abstract NOTCH1 receptor signaling plays a central role in T-cell lineage specification and in supporting the growth and proliferation of immature T-cell progenitors in the thymus during lymphoid development. In T-cell acute lymphoblastic leukemia (T-ALL), a tumor resulting from the malignant transformation of T-cell progenitors, aberrant and constitutively active NOTCH1 signaling triggered by activating mutations in the *NOTCH1* gene contributes to oncogenic transformation and is a hallmark of this disease. Most notably, small molecule γ-secretase inhibitors (GSIs) can effectively block NOTCH1 signaling in T-ALL, and could be exploited as a targeted therapy in this disease. In addition, a number of emerging anti-NOTCH therapeutic strategies including anti-NOTCH1 inhibitory antibodies, small peptide inhibitors of NOTCH signaling and combination therapies with GSIs and glucocorticoids, have recently been proposed. Finally, the identification of *NOTCH1* mutations in solid tumors and chronic lymphocytic leukemias has increased even further the clinical relevance of NOTCH signaling as a therapeutic target in human cancer. Here we review our current understanding of NOTCH1-induced transformation, the mechanisms of action of oncogenic NOTCH1 in T-ALL and the therapeutic and prognostic implications of *NOTCH1* mutations in T-ALL.

G. Tzoneva
Institute for Cancer Genetics and Graduate Program in Pathobiology and Molecular Medicine, Columbia University Medical Center, New York 10032, USA

A. A. Ferrando (✉)
Departments of Pediatrics and Pathology and Institute for Cancer Genetics, Columbia University Medical Center, New York 10032, USA
e-mail: af2196@columbia.edu

Current Topics in Microbiology and Immunology (2012) 360: 163–182
DOI: 10.1007/82_2012_232

Published Online: 7 June 2012

Contents

1 Introduction

T-cell acute lymphoblastic leukemia (T-ALL) is an aggressive hematologic tumor characterized by bone marrow infiltration by leukemia lymphoblasts expressing immature T-cell markers. T-ALL accounts for 10–15 % of pediatric and 25 % of adult acute lymphoblastic leukemias. These patients frequently show large tumor burdens at diagnosis with very high white blood cell counts, mediastinal masses, and pleural effusions. In addition, T-ALL patients have a higher risk of meningeal central nervous system infiltration at diagnosis. Notably, in the early days of combination chemotherapy, when treatment of B-precursor ALL with combination chemotherapy started to make significant progress in survival, clinical results in T-ALL lagged behind due to an exceedingly high rate of relapse. The dismal prognosis of T-ALL with cure rates under 10 % resulted in the introduction of intensified chemotherapy protocols for this disease. Since then, the outcomes of T-ALL patients have significantly improved, with current 5-year relapse-free survival rates over 75 % in children and over 50 % in adults. However, the prognosis of T-ALL patients who present with primary resistant disease or relapse after the induction of complete remission with chemotherapy remains very poor, highlighting the need to develop more specific and highly effective antileukemic drugs. Our understanding of the molecular basis of T-ALL has improved dramatically over the last decade. Originally considered a single disease, gene expression profiling studies have revealed a marked molecular heterogeneity in this disease. Thus, T-ALL comprises distinct molecular groups with unique gene expression signatures, a different arrest in T-cell development and different responses to therapy (Aifantis et al. 2008; Ferrando et al. 2002).

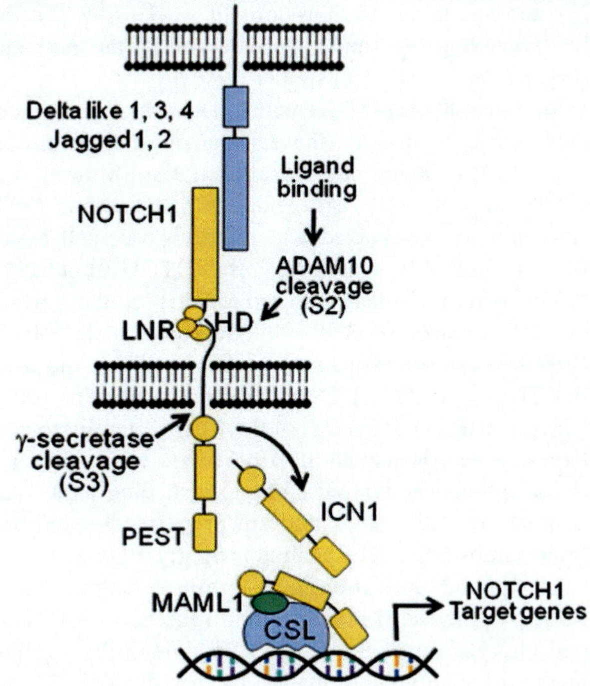

Fig. 1 The NOTCH1 signaling pathway. Schematic representation of NOTCH1 activation. Interaction of the NOTCH1 receptor with Delta-like and Jagged ligands expressed on the surface of a neighbor cells triggers the proteolytic cleavage of the receptor, first by an ADAM metalloprotease (S2 cleavage) and subsequently by the γ-secretase complex (S3 cleavage), which releases the intracellular domains of NOTCH1 (ICN1) from the membrane. ICN1 translocates to the nucleus and interacts with DNA via the RBPJ/CSL DNA binding protein and recruits the MAML1 coactivator to activate the expression of NOTCH1 target genes. Reproduced with permission from (Ferrando 2009)

2 The NOTCH Signaling Pathway

NOTCH receptors (NOTCH1–4) function as ligand-activated transcription factors to directly transduce extracellular signals into changes in gene expression in the nucleus. Each of these receptors is a class I transmembrane glycoprotein expressed at the cell surface as a heterodimer of an N-terminal extracellular (N^{EC}) fragment and a C-terminal transmembrane-intracellular subunit (N^{TM}) generated by proteolytic cleavage of a proNOTCH precursor polypeptide by a furin-like protease in the trans-Golgi network (Fig. 1) (Blaumueller et al. 1997; Logeat et al. 1998). The extracellular (N^{EC}) fragment of the receptor is responsible for interaction with ligands from the Delta and Serrate family (Delta-like 1, 3, and 4; and Jagged 1 and 2) (Rebay et al. 1991). In the absence of ligand, the N^{EC} is closely associated with the C-terminal transmembrane and cytoplasmic fragment of the receptor (N^{TM}) via a heterodimerization domain (HD). Importantly, three LNR repeats in the N^{EC} fragment fold over the HD domain, stabilizing, and shielding the HD domain in a molecular lock that prevents NOTCH activation in the absence of ligand (Rand et al. 2000; Sanchez-Irizarry et al. 2004). The N^{TM} subunit contains a single-pass transmembrane domain and a cytoplasmic region which functions as a ligand-activated transcription factor. Importantly, the C-terminal part of the N^{TM} contains a PEST [proline (P), glutamic acid (E), serine

(S), and threonine (T) rich] domain, responsible for terminating NOTCH signaling by targeted proteasomal degradation of the activated receptor in the nucleus (Fryer et al. 2004; Rechsteiner 1988).

Interaction of NOTCH with a Delta-like or Jagged ligand on the surface of a neighboring cell triggers the cleavage of the extracellular fragment of the receptor by the ADAM10 membrane metalloprotease (Mumm et al. 2000; van Tetering et al. 2009). This S2 cleavage generates a truncated NOTCH intermediate which is subsequently recognized and rapidly processed by the γ-secretase complex, an aspartyl protease which cleaves NOTCH at a site within the cell membrane (S3 cleavage) to release the active, intracellular portion of NOTCH (ICN) into the cytosol (De Strooper et al. 1999; Schroeter et al. 1998; Struhl and Greenwald 1999). ICN then rapidly translocates to the nucleus and associates with the RBPJ/CSL (CBF1/Su(H)/LAG-1) DNA binding protein to form a transcriptional complex which recruits coactivators of the MAML family to activate expression of NOTCH target genes (Hsieh et al. 1996; Wu et al. 2000). The core of the NOTCH transcriptional activation complex can bind as a monomer, however it can also dimerize on DNA sites that are properly oriented and spaced (Liu et al. 2010). Importantly, NOTCH signaling is tightly regulated by actively coupling ICN-driven transcriptional activation to ICN protein degradation. Specifically, recruitment of RNA polymerase II holoenzyme to the ICN-MAML-RBPJ/CSL complex triggers phosphorylation of the ICN PEST domain by cyclin-dependent kinase 8, which marks it for polyubiquitination by the FBXW7-SCF ubiquitin ligase complex and subsequent proteasomal degradation in the nucleus (Fryer et al. 2004).

3 Mechanisms of Aberrant NOTCH1 Activation in T-ALL

The initial discovery of *NOTCH1* and its aberrant activation in T-ALL came from analysis of the rare t(7;9)(q34;q34.3) chromosomal translocation recurrently found in a small fraction of T-ALL cases. This rearrangement juxtaposed a truncated *NOTCH1* gene next to the *TCRB* (T-cell receptor beta) locus, leading to aberrant expression of truncated polypeptides encoded by the derivative *NOTCH1* allele (Ellisen et al. 1991; Palomero et al. 2006a). These N-terminally truncated forms of NOTCH1 were either membrane-bound receptors constitutively processed by γ-secretase, or intracellular proteins without a transmembrane domain that did not even require γ-secretase processing for activation. The exact location of the translocation breakpoint and the translation initiation site determined which of these truncated NOTCH1 proteins would be formed (Ellisen et al. 1991; Palomero et al. 2006a).

Subsequent mouse studies demonstrated that expression of similar alleles of constitutively active Notch1 in hematopoietic progenitors leads to potent induction of murine T-ALL (Pear et al. 1996), and that Notch1 activation is a common event in murine T-cell neoplasms induced by retroviral mutagenesis (Girard et al. 1996). The specific oncotropic role of Notch1 for T-cell progenitors is consistent with the

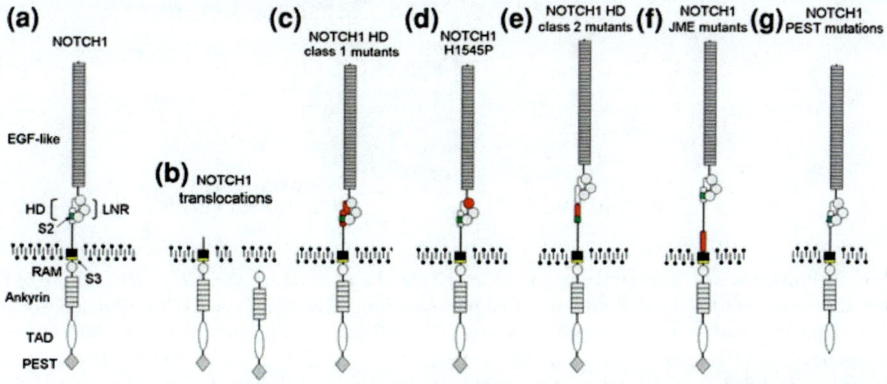

Fig. 2 Oncogenic forms of NOTCH1 in T-ALL. Aberrant activation of NOTCH signaling can be triggered by mutations in the NOTCH1 gene. **a** Structure of the wild type NOTCH1 receptor. Functional domains of NOTCH1 are annotated. EGF-like: EGF-like repeats. HD: heterodimerization domain. LNR: LNR repeats. RAM: RAM domain. Ankyrin: ankyrin repeats. TAD: transactivation domain. PEST: PEST domain. S2: metalloprotease cleavage site (*green*). S3: γ-secretase cleavage site (*yellow*). **b** Translocations of *NOTCH1* to the *TCR* loci induce the expression of truncated forms of NOTCH1. **c** *NOTCH1 HD class1* mutations destabilize the structure of the HD-LNR repeats responsible for maintaining the receptor in resting configuration. **d** The *NOTCH1 H1545P* mutation impairs the protection of the S2 cleavage site by the HD-LNR repeat complex. **e** *NOTCH1 HD class2* mutations displace the S2 metalloprotease cleavage site outside the HD-LNR complex. **f** *NOTCH1 JME* alleles increase the separation of the HD-LNR repeat complex from the membrane. **g** *NOTCH1 ΔPEST* mutations delete the C-terminal part of the receptor and impairing the degradation of activated NOTCH1 in the nucleus. Sequences altered by the different *NOTCH1* mutations are highlighted in red. Reproduced with permission from (Ferrando 2009)

strict requirement and lineage instructive role of Notch1 for T-cell development from hematopoietic stem cells (Pui et al. 1999; Radtke et al. 1999). However, since aberrant NOTCH1 activation seemed to be restricted to only a handful of human tumors harboring the t(7;9) (q34;q34.3) translocation, the significancc of these findings for the pathogenesis of T-ALL remained elusive.

This radically changed with the discovery of mutations in *NOTCH1* in over 60 % of human T-ALLs (Weng et al. 2004). *NOTCH1* mutations in T-ALL are activating alleles resulting in abnormally high levels of NOTCH1 signaling. The most frequent mutational hotspots in *NOTCH1*, found in approximately 40 % of human T-ALLs, are exons 26 and 27, encoding the N-terminal and C-terminal components of the heterodimerization domain, respectively (Fig. 2 and 3) (Weng et al. 2004). The structural basis for the mechanism of action of these mutations became understood with the elucidation of the crystal structure of the Notch1 LNR-HD complex, which shows the two HD subunits to be closely associated and held together by the LNR repeats, which fold over to stabilize the HD interaction (Gordon et al. 2009). Furthermore, the third LNR domain acts to directly shield the S2 cleavage site from extracellular metalloproteases (Gordon et al. 2009). Most *NOTCH1 HD* mutations (class 1, HD1) are typically

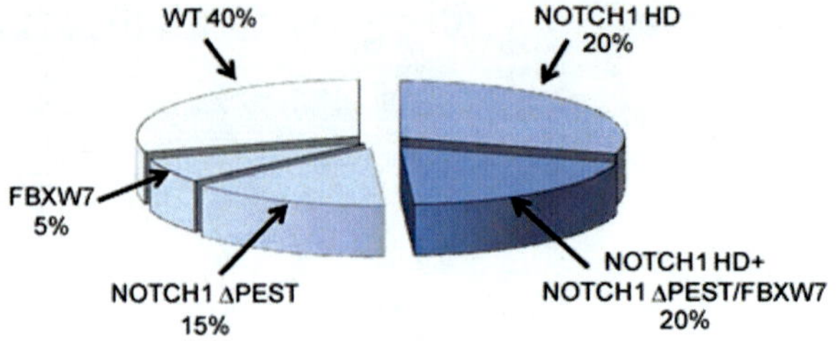

Fig. 3 Prevalence of activating mutations in NOTCH1 in T-ALL. *NOTCH1 HD* and *ΔPEST* mutations account for the majority of activating mutations in NOTCH1 in T-ALL. An additional 1 % of patients harbor translocations involving *NOTCH1* and the *TCR* loci. *NOTCH1 JME* alleles are found in 3 % of all T-ALLs. In total about 15 % of T-ALL cases harbor deletions or mutations in *FBXW7*, which impair the degradation of activated NOTCH1 in the nucleus and are functionally related to *NOTCH1 ΔPEST* mutant alleles. Reproduced with permission from (Ferrando 2009)

single amino acid substitutions and small in-frame deletions and insertions (Fig. 2), which compromise the stability of the LNR-HD complex, leading to ligand-hypersensitivity or ligand-independent NOTCH1 activation (Malecki et al. 2006). Class 2 HD mutations (HD2) are longer insertions located at the distal part of the HD domain which expose the S2 cleavage site to extracellular metallo-protease activity by shifting it away from the protective cover of the LNR-HD complex (Fig. 2) (Malecki et al. 2006). Similarly, the *NOTCH1 H1545P* mutation, which is located in the LNR repeat region, acts by loosening the protective LNR shield from the S2 cleavage site (Gordon et al. 2009) (Fig. 2).

Juxtamembrane Expansion Mutants (*JME NOTCH1*) are internal tandem duplications in the 3′ end of intron 27 and/or the proximal region of exon 28, and are yet another group of mutations which augment NOTCH1 activation at the cell membrane (Sulis et al. 2008). *JME NOTCH1* mutations do not alter the structure of the LNR-HD complex but displace the complex away from the membrane (Fig. 2), which results in very high levels of NOTCH1 processing and signaling.

Another *NOTCH1* mutational hotspot in T-ALL is the PEST domain, with 20–25 % of T-ALLs exhibiting *NOTCH1 ΔPEST* mutations (Fig. 3). In these alleles the NOTCH1 PEST domain is truncated or lost due to frameshift or nonsense nucleotide substitutions, which generate premature stop codons in the C-terminal portion of the receptor. *ΔPEST* mutations impair FBXW7-mediated targeting of ICN1 for proteasomal degradation, leading to increased levels of activated NOTCH1 (Thompson et al. 2007; Weng et al. 2004). In addition, 15 % of T-ALL cases harbor mutations or deletions in *FBXW7* (Asnafi et al. 2009; O'Neil et al. 2007; Thompson et al. 2007), which cluster in three arginine residues critical for its interaction with ICN1 (Fig. 3). FBXW7 mutant proteins have impaired substrate recognition and mimic the effect of *NOTCH1 ΔPEST* mutations by promoting increased stability of activated NOTCH1 (O'Neil et al. 2007; Thompson et al. 2007). FBXW7 is, however, also responsible for tagging additional oncoproteins such as c-MYC, JUN, mTOR, and Cyclin E for degradation (Minella and Clurman 2005). Therefore mutational inactivation of FBXW7 is likely to have much broader biological consequences on T-ALL pathogenesis than *NOTCH1 ΔPEST* mutations, possibly leading to increased cell metabolism, cell growth, and cell cycle progression.

Another important observation is that about 15 % of T-ALLs harbor *NOTCH1 HD* and *ΔPEST* mutations in *cis*, i.e., in the same transcript (Fig. 3), (Weng et al. 2004), while 5 % of cases present with a *NOTCH1 HD* and a *FBXW7* mutation (Fig. 3), (O'Neil et al. 2007; Thompson et al. 2007). The occurrence of ligand-independent NOTCH activation at the membrane and impaired degradation of ICN1 in the nucleus in these leukemias leads to exceedingly high levels of NOTCH1 activity (Thompson et al. 2007; Weng et al. 2004).

Recent studies in mice have shown that the different human *NOTCH1* gain-of-function mutations have marked differences in their potential to induce T-cell leukemia (Chiang et al. 2008). Expression in hematopoietic precursors of less common, stronger mutant *NOTCH1* alleles such as *HD1-ΔPEST* or *HD2*, which produce high levels of active NOTCH1, was sufficient to induce T-cell transformation. In contrast, more common, weaker mutant alleles like *HD1* or *ΔPEST* could drive ectopic T-cell development but were not sufficient by themselves to induce leukemia in mice. However, expression of these weaker *NOTCH1* alleles in hematopoietic progenitors expressing constitutively active K-ras effectively triggered the accelerated development of T-ALL tumors (Chiang et al. 2008). The heterogeneous nature of mutant *NOTCH1* alleles suggests that different *NOTCH1* mutations may serve different roles in T-ALL, all the while contributing to increasing the levels of NOTCH1 signaling. In this context strong *NOTCH1* alleles may act as primary drivers of T-cell transformation while weaker alleles could play an adjuvant role and work in concert with other oncogenic lesions. The identification of a *NOTCH1 HD* mutation in the neonatal blood spots of a patient who developed T-ALL at 6 years of age shows that *NOTCH1* mutations may occur as initiating events prenatally (Eguchi-Ishimae et al. 2008). On the other hand, subclonal *NOTCH1* mutations are frequently found at diagnosis of T-ALL, with relapsed samples displaying a change to wild type *NOTCH1* or containing a different *NOTCH1* mutant clone (Mansour et al. 2007). Therefore, *NOTCH1* mutations may be initiating or secondary events in T-ALL progression. How these functional differences may affect the response to anti-NOTCH1 therapies in the clinic remains to be elucidated.

Finally, two new mechanisms of Notch1 activation have been recently described in mouse models of T-ALL. First, alternative promoter usage with activation of an intragenic promoter associated with expression of truncated Notch1 receptors with ligand-independent activity and intragenic deletions in *Notch1*, has been described in Ikaros1-deficient mice (Gomez-del Arco et al. 2010). In addition, intragenic deletions in the *Notch1* gene resulting in high levels of Notch1 activation and expression of 5′ truncated alternatively spliced forms of NOTCH1 have been reported as common oncogenic alterations in mouse models of T-ALL (Ashworth et al. 2010; Jeannet et al. 2010; Tsuji et al. 2003).

4 Prognostic Value of NOTCH1 Mutations in T-ALL

A number of clinical studies have analyzed the prognostic significance of NOTCH mutations, and while associations between NOTCH1 mutational status and clinical outcome have not been consistent, several analyses have shown an association of *NOTCH1* mutations with better prognosis in T-ALL (Ferrando 2010). An analysis of 157 pediatric T-ALL patients from the ALL-Berlin-Frankfurt-Munster (BFM) 2000 study found that *NOTCH1* mutations were associated with better early treatment response and improved relapse-free survival (Breit et al. 2006). Similarly, an examination of the *NOTCH1* and *FBXW7* mutational status of 55 T-ALL and 14 T-cell lymphoblastic lymphoma patients treated in the Japan Association of Childhood Leukemia Study (JACLS) protocols ALL-97 and NHL-98 showed that *NOTCH1* and/or *FBXW7* mutations correlated with better event-free and overall survival (Park et al. 2009). *NOTCH1* and/or *FBXW7* mutations were also found to be positive prognostic factors in adult patients treated in the Lymphoblastic Acute Leukemia in Adults (LALA)-94 ($n = 87$) and the GRAALL-2003 ($n = 54$) trials (Asnafi et al. 2009).

In contrast, analysis of 72 pediatric T-ALL patients treated in the Dutch Childhood Oncology Group (DCOG) protocols ALL-7, ALL-8, or ALL-9 showed no difference in prognosis for patients whose leukemias harbored activating mutations in *NOTCH1* (van Grotel et al. 2006), and an analysis of 88 adult T-ALL patients treated according to the MRC UKALLXII/ECOGE2993 protocol found no association between *NOTCH1* and/or *FBXW7* mutations and clinical outcome (Mansour et al. 2009).

In an effort to settle some of these discrepancies, three later studies reexamined the prognostic significance of *NOTCH1* and *FBXW7* mutations. Analysis on an extended cohort of 301 patients treated with the ALL-BFM protocol confirmed the overall favorable effect of activating *NOTCH1* mutations on long-term outcome, but also determined that this positive prognosis is restricted to patients with rapid early treatment response (Kox et al. 2010). Inactivating *FBXW7* mutations were found to correlate with a highly favorable prednisone response, but had no appreciable association with long-term outcome and did not synergize with *NOTCH1* mutations (Kox et al. 2010). Another study on a series of 134 pediatric patients from the EORTC-CLG 58881 and 58951 protocols reported that *NOTCH1* and *FBXW7* mutations associate with improved early chemotherapeutic response and lower minimal residual disease (MRD) levels (Clappier et al. 2010). This, however, did not translate into improved long-term outcome and, more strikingly, *NOTCH1/FBXW7* mutations correlated with worse outcome in patients with very high MRD levels (Clappier et al. 2010). Finally, Zuurbier et al. describe a retrospective analysis performed on 72 patient samples from the DCOG ALL7/8 and ALL9 protocols, and 74 cases from the COALL-97 clinical trial (Zuurbier et al. 2010). Using reverse-phase protein microarrays to directly measure levels of activated NOTCH1, as well as microarray expression analysis of NOTCH target genes, the study confirmed that *NOTCH1* and *FBXW7* mutations are indeed associated with increased intracellular NOTCH1 activity. Similarly to the EORTC-CLG studies, the presence of *NOTCH1/FBXW7* mutations correlated with a better initial prednisone response, but had no effect on long-term survival (Zuurbier et al. 2010). Overall these results suggest that *NOTCH1* and *FBXW7* mutations may be associated with increased response to treatment, but that differences in therapy may affect the ultimate impact of these mutations on prognosis.

5 Genes and Pathways Controlled by NOTCH1 in T-cell Transformation

The past few years have seen a strong effort to identify direct NOTCH1 target genes and to decipher the genetic programs downstream of NOTCH1 in T-ALL. Integrative analysis of NOTCH1 direct target genes and gene expression programs controlled by NOTCH1 have shown that NOTCH1 promotes leukemic cell growth via direct transcriptional upregulation of multiple anabolic genes involved in ribosome biosynthesis, protein translation, and nucleotide, and amino acid metabolism (Palomero et al. 2006b; Weng et al. 2006). In addition, NOTCH1 also has an indirect, yet equally significant, role in promoting cell growth and metabolism by directly inducing the expression of the *MYC* oncogene (Palomero et al. 2006b; Sharma et al. 2006; Weng et al. 2006). Notably, MYC further promotes the expression of anabolic genes (Margolin et al. 2009; Palomero et al. 2006b) and shares many direct target genes with NOTCH1 (Palomero et al. 2006b). Still, NOTCH1 seems to be oncogenic dominant over MYC (Demarest et al. 2011). This NOTCH-MYC feed-forward-loop transcriptional regulatory network positions NOTCH1 as a critical driver of cell growth and anabolism in T-ALL.

Yet another crucial mechanism by which NOTCH1 acts to regulate cell growth is by activating the PI3K-AKT-mTOR pathway, which mediates multiple growth factor-driven cellular responses involved in cell growth, proliferation, and survival. Initial studies in T-cell precursors demonstrated that NOTCH signaling regulates cell size, glucose uptake, and glycolysis via PI3K-AKT activation (Ciofani and Zuniga-Pflucker 2005). Mechanistically, it was shown early on that NOTCH1 can activate AKT via the $p56^{lck}$ tyrosine kinase in T-cells (Sade et al. 2004), while further studies in T-cell progenitors and T-ALL lymphoblasts showed that the *PTEN* tumor suppressor gene—a strong inhibitor of the PI3 K-AKT pathway—is downregulated by the HES1 transcriptional repressor, one of the most prominent NOTCH1 direct target genes (Palomero et al. 2007). Consistently, T-ALL cell lines treated with NOTCH1 inhibitors show a marked decrease in the phosphorylation of mTOR targets, as demonstrated by phosphoproteomic analysis (Chan et al. 2007). In addition, NOTCH signaling upregulates the expression of important cytokine and growth factor receptors signaling upstream of the PI3K pathway. Most prominently, NOTCH1 regulates the expression of the pre TCR alpha (PTCRA), which in association with TCRB provides the signal for clonal expansion of T-cell progenitor cells as they transition from the early double negative to the more mature double positive stages of thymocyte development (von Boehmer et al. 1998). Similarly, NOTCH1 directly controls the expression of the interleukin 7 receptor alpha chain (IL7R) in T-cell progenitors and T-ALL (Gonzalez-Garcia et al. 2009). IL7 signaling is strictly required for the proliferation and survival of T-cell progenitors, and the role of the IL7R in T-cell transformation is supported by the identification of oncogenic *IL7R* gain-of-function mutations in 10 % of T-ALLs (Porcu et al. 2012; Shochat et al. 2011; Zenatti et al. 2011; Zhang et al. 2012). Similarly, the expression of IGF1R was recently shown to be important for T-ALL cell growth and leukemia-initiating activity in vivo, and to be directly controlled by NOTCH1 (Medyouf et al. 2011).

In addition to promoting anabolic cell growth, oncogenic NOTCH1 signaling has been shown to directly promote proliferation via increased G1/S cell cycle progression in T-ALL (Dohda et al. 2007; Joshi et al. 2009; Rao et al. 2009). The *CCND3*, *CDK4*, and *CDK6* cell cycle genes are upregulated by NOTCH1 activation in T-ALL, and *CCND3* is a direct NOTCH1 target gene (Joshi et al. 2009). Consistently, inhibition of NOTCH signaling in T-ALL cell lines induces upregulation of the cyclin-dependent kinase inhibitors *CDKN2D (p19/INK4d)* and *CDKN1B (p27/Kip1)* (Rao et al. 2009). Moreover, Notch1 can promote premature entry of hematopoietic progenitors into S phase by inducing transcription of the S phase kinase-associated protein 2 (SKP2), which mediates the proteasomal degradation of the CDKN1B (p27/Kip1) and CDKN1A (p21/Cip1) cell cycle inhibitors (Dohda et al. 2007).

NOTCH1 may also be a regulator of p53 activity (Beverly et al. 2005) and can promote cell migration via upregulation of RHOU (Bhavsar et al. 2012) and the CCR5 and CCR9 chemokine receptors (Mirandola et al. 2012). Importantly, NOTCH1 signaling also upregulates the expression of CCR7 in T-ALL cells and provides an essential adhesion signal that facilitates leukemia infiltration in the central nervous system (Buonamici et al. 2009). In addition, NOTCH1 signaling enhances NFKB activity in T-ALL by increasing NFKB expression (Thompson et al. 2007), promoting NFKB nuclear retention (Shin et al. 2006), and via activation of IκB kinase (IKK) (Song et al. 2008). Moreover, Notch1 promotes IKK activation via Hes1-mediated repression of the deubiquitinase CYLD, a negative IKK complex regulator (Espinosa et al. 2010), whose repression is an essential event downstream of Notch1 in T-ALL (D'Altri et al. 2011).

Finally, conditional inactivation of Hes1 in mice impairs NOTCH1 induced transformation (Wendorff et al. 2010), which suggests an important role for this transcriptional repressor in mediating the oncogenic program activated by NOTCH1 in T-ALL.

6 NOTCH1 as a Therapeutic Target in T-ALL

The presence of NOTCH1 mutations in T-ALL has generated great interest in the development of anti-NOTCH1 targeted therapies for the treatment of this disease. To date, a large part of this effort has focused on inhibiting the γ-secretase complex, as proteolytic processing of the receptor is required for the activity of all mutant forms of NOTCH1. Importantly, the γ-secretase complex has been implicated in the pathogenesis of Alzheimer's disease, where it is responsible for the pathologic processing of the amyloid precursor protein (APP) into amyloidogenic Aβ peptides (Selkoe and Kopan 2003). Notably, small molecule γ-secretase inhibitors (GSIs) developed for the treatment of Alzheimer's disease can also effectively block γ-secretase activation of NOTCH1 (Evin et al. 2006). Indeed, GSI treatment of T-ALL cell lines harboring activating mutations in *NOTCH1* induced rapid clearance of intracellular activated NOTCH1 protein and transcriptional downregulation of NOTCH1 target genes (Palomero et al. 2006a; Palomero et al. 2006b; Weng et al. 2004), which was accompanied by a G1 cell cycle arrest and a decrease in cell size in some cell lines (Palomero et al. 2006a; Weng et al. 2004). Notably, ex vivo inhibition of NOTCH signaling in primary T-ALL cells effectively prevents these tumors from engrafting in immunodeficient mice, thus suggesting that inhibition of NOTCH1 signaling could reduce the self-renewal capacity of T-ALL lymphoblasts (Armstrong et al. 2009; Tatarek et al. 2011).

Following these results, a phase I, open label, nonrandomized clinical trial, Dana-Farber Cancer Institute 04-390, was started in order to test the activity of MK-0752, an oral GSI developed by Merck for the treatment of relapsed T-ALL (Deangelo et al. 2006). In this trial, GSI treatment led to a downregulation of *NOTCH1* target genes in T-cell lymphoblasts, however a significant number of patients developed marked fatigue and dose-limiting gastrointestinal toxicity, most likely attributed to the inhibition of NOTCH signaling in the gut, and there were no objective responses (Deangelo et al. 2006).

These disappointing results highlight some of the challenges facing the development of GSIs as anti-NOTCH1 therapies in the treatment of T-ALL. First, GSIs were actually found to have only limited antileukemic activity against human T-ALL cell lines, exerting primarily a cytostatic effect with minimal or no apoptosis (Palomero et al. 2006a; Weng et al. 2004). The weak effect of GSIs on human T-ALLs appears to be in stark contrast to its effect on mouse T-cell leukemias induced by constitutive NOTCH1 activation, where a robust antileukemic and apoptotic response is induced by GSI inhibition of NOTCH signaling (Chiang et al. 2008; Cullion et al. 2009). The established role of NOTCH signaling in promoting cell growth and metabolism suggests that additional mutations in oncogenic pathways promoting cell growth could serve to ameliorate the effects of GSI-mediated NOTCH inhibition in human T-ALL lymphoblasts. Comparison of GSI sensitive and resistant T-ALL cell lines demonstrated that GSI resistance can be induced by mutational loss of *PTEN* which leads to constitutive activation of the PI3K-AKT-mTOR pathway, bypassing the requirement for NOTCH1 signaling for continuous growth and metabolism (Palomero et al. 2007). However, *Pten* null primary leukemias can still respond to some extent to NOTCH inhibition (Medyouf et al. 2010). Forced expression of *c-MYC* has also been shown to overcome the cell cycle arrest induced by NOTCH1 inhibition in some T-ALL cell lines (Weng et al. 2006). Finally, GSI resistant T-ALL cell lines are found to harbor *FBXW7* mutations more frequently than GSI sensitive cell lines, suggesting that increased levels of c-MYC, JUN, cyclin E, mTOR, and other oncoproteins degraded by FBXW7 may lead to reduced sensitivity to NOTCH1 inactivation (O'Neil et al. 2007; Thompson et al. 2007).

Thus *PTEN* loss, *MYC* overexpression, and *FBXW7* mutations could play a major role in T-ALL resistance to NOTCH1 inhibition, so their effect on clinical responses to GSI treatment should be investigated.

A second major challenge to GSI treatment is the on-target gastrointestinal toxicity observed in T-ALL patients treated with MK-0752. NOTCH signaling is a crucial regulator of lineage specification in many tissues, influencing the lineage commitment decisions of multipotent precursor cells. The gut is home to both NOTCH1 and NOTCH2 receptors, which are expressed on intestinal stem cells and play essential roles in the homeostasis of intestinal progenitors (Riccio et al. 2008). Blocking Notch signaling by conditional targeting of *RBPJ/CSL* (van Es et al. 2005) or via conditional knockout of *Notch1* and *Notch2* (Riccio et al. 2008) in the gut blocks cell proliferation in the intestinal crypt and leads to secretory cell hyperplasia at the expense of absorptive enterocytes. Inhibition of NOTCH signaling with a GSI has a very similar effect, inducing secretory metaplasia in the intestine with loss of the proliferative compartment and accumulation of mucus-secreting goblet cells (Real et al. 2009; Wei et al. 2010). Mechanistically, loss of *Hes1* inhibits the expression of *MATH1* and *Klf4*, two transcription factors required for goblet cell production in the gut (Katz et al. 2002; Yang et al. 2001). Notably, Hes1 loss mimics in part the effects of NOTCH inactivation in the gut (Jensen et al. 2000) and Math1 knockout animals are resistant to GSI-induced secretory metaplasia (Kim and Shivdasani 2011). Overall, these results support that suppression of a Notch-Hes1-Math1/Klf4 regulatory axis upon NOTCH inhibition leads to goblet cell accumulation primarily via uncontrolled expression of Math1.

7 Combination Therapies With GSIs in the Treatment of T-ALL

Development of successful anti-NOTCH therapies for T-ALL will have to tackle the challenges of limited antileukemic activity and extensive gut toxicity which are observed with GSIs. It is possible that new, more active GSIs could overcome the resistance of tumor cells to NOTCH inhibition, while intermittent treatment schedules can increase tolerance to GSI therapy (Cerchietti et al. 2010; Tammam et al. 2009; Wei et al. 2010). In addition, combination treatments offer the possibility of great improvements in both efficacy and lowering of gut toxicity. Thus, combinations of GSIs with cyclin-dependent kinase inhibitors (Rao et al. 2009), drugs targeting NFKB signaling (Thompson et al. 2007), or small molecule inhibitors of the PI3K-AKT-mTOR pathway (Chan et al. 2007; Cullion et al. 2009; Palomero et al. 2007) could lead to greater antileukemic activity. Nonetheless, a more rational and direct approach to designing GSI combination therapies for T-ALL would be to use GSIs as a supplement to current standard T-ALL drug treatments. Constitutive activation of NOTCH1 signaling can protect developing thymocytes from glucocorticoid-induced apoptosis (Deftos et al. 1998), suggesting that GSIs and glucocorticoids could have a synergistic antileukemic effect in

T-ALL. Indeed, inhibition of NOTCH signaling with a GSI effectively sensitizes glucocorticoid-resistant T-ALL cell lines and primary patient samples to glucocorticoid-induced apoptosis (Real et al. 2009). Mechanistically, NOTCH1-HES1 downregulate a positive feedback loop in glucocorticoid-induced apoptosis mediated by transcriptional glucocorticoid receptor auto-upregulation (Real et al. 2009). An interesting observation is that glucocorticoid treatment downregulates NOTCH signaling via serum and glucocorticoid-inducible kinase 1 activation of FBXW7 (Mo et al. 2011), suggesting a dual interaction between the NOTCH1 and the glucocorticoid signaling pathways. Most notably, in vivo administration of GSI–glucocorticoid combination treatments in a mouse model of glucocorticoid-resistant T-ALL showed that glucocorticoids have a direct protective effect against GSI-induced intestinal toxicity (Real et al. 2009). Hence, these studies strongly suggest that a combination treatment of glucocorticoids and GSIs could enhance the antileukemic effects of these anti-NOTCH drugs and reduce intestinal toxicity associated with systemic inhibition of NOTCH signaling.

8 Novel Anti-NOTCH Therapies

Transcription factors are notoriously challenging drug targets and as such most efforts to inhibit NOTCH signaling have focused on blocking the receptor at the membrane. However, MAML1 dominant negative peptides can interfere with the assembly of the ICN1-RBPJ-MAML1 transcriptional complexes in the nucleus and effectively abrogate NOTCH signaling (Weng et al. 2003). This principle was exploited by Moellering et al. to develop a synthetic, cell-permeable, stabilized alpha-helical peptide (SAHM1) which binds with high affinity to a protein–protein interface on the NOTCH-CSL transactivation complex, effectively blocking MAML1 recruitment and preventing NOTCH-mediated transcription (Moellering et al. 2009). Treatment with SAHM1 of human T-ALL cell lines and a mouse model of NOTCH1-driven T-ALL resulted in strong, NOTCH-specific inhibition of cell proliferation, and leukemia progression (Moellering et al. 2009).

However, small peptide inhibitors of the NOTCH transcriptional complex will still block all four NOTCH receptors, with potential side effects resulting from the simultaneous inhibition of NOTCH1 and 2 in the intestine. In contrast, inhibitory antibodies with selective specificity against NOTCH1 could potentially selectively block NOTCH1 in leukemic clones while sparing the other three NOTCH receptors. In this regard, anti-NOTCH1 antibodies capable of stabilizing the HD-LNR region in its quiescent state can effectively block the activation of wild type NOTCH1 and at least some of the leukemia-associated NOTCH1 receptor mutants (Aste-Amezaga et al. 2010; Wu et al. 2010).

Finally, targeted inhibition of ADAM10 (Bozkulak and Weinmaster 2009; Sulis et al. 2011) or other surface proteins implicated in NOTCH1 processing and activation such as TSPAN33 and TSPAN5 (Dunn et al. 2010) may provide additional avenues for the development of anti-NOTCH therapies in T-ALL.

9 Concluding Remarks and Future Directions

Recent progress in the characterization of the molecular mechanisms that mediate the oncogenic activity of NOTCH signaling in T-ALL have opened new possibilities for the development of anti-NOTCH therapies. In addition, activating mutations in *NOTCH1* have been found in 10 % of non-small-cell lung carcinomas (Westhoff et al. 2009) and chronic lymphocytic leukemias (CLL) (Puente et al. 2011). Most notably, *NOTCH1* CLL mutations are associated with poor prognosis and are particularly frequent in cases showing resistance to chemotherapy and upon transformation to high grade lymphoma. However, loss of function mutations in the NOTCH pathway have also been found in cutaneous, lung, and head and neck squamous cell carcinomas, (Wang et al. 2011) and in myeloproliferative disease (Klinakis et al. 2011), pointing to a tumor suppressor role for NOTCH signaling in epithelial cells and myeloid progenitors. These warning signals suggest that anti-NOTCH therapies should be limited in their duration in order to avoid the development of tumors resulting from prolonged suppression of NOTCH signaling in these tissues. Still, the progress in our understanding of the mechanisms of transformation downstream of NOTCH have created a renewed interest in the NOTCH signaling pathway as a therapeutic target in human cancer and generated momentum for the design of new clinical trials testing the efficacy and safety of anti-NOTCH therapies in the clinic.

References

Aifantis I, Raetz E, Buonamici S (2008) Molecular pathogenesis of T-cell leukaemia and lymphoma. Nat Rev Immunol 8:380–390

Armstrong F, de la Grange PB, Gerby B, Rouyez MC, Calvo J, Fontenay M, Boissel N, Dombret H, Baruchel A, Landman-Parker J, Romeo PH, Ballerini P, Pflumio F (2009) NOTCH is a key regulator of human T-cell acute leukemia initiating cell activity. Blood 113:1730–1740

Ashworth TD, Pear WS, Chiang MY, Blacklow SC, Mastio J, Xu L, Kelliher M, Kastner P, Chan S, Aster JC (2010) Deletion-based mechanisms of notch1 activation in T-ALL: key roles for RAG recombinase and a conserved internal translational start site in notch1. Blood 116:5455–5464

Asnafi V, Buzyn A, Le Noir S, Baleydier F, Simon A, Beldjord K, Reman O, Witz F, Fagot T, Tavernier E, Turlure P, Leguay T, Huguet F, Vernant JP, Daniel F, Bene MC, Ifrah N, Thomas X, Dombret H, Macintyre E (2009) NOTCH1/FBXW7 mutation identifies a large subgroup with favorable outcome in adult T-cell acute lymphoblastic leukemia (T-ALL): a Group for Research on adult acute lymphoblastic leukemia (GRAALL) study. Blood 113:3918–3924

Aste-Amezaga M, Zhang N, Lineberger JE, Arnold BA, Toner TJ, Gu M, Huang L, Vitelli S, Vo KT, Haytko P, Zhao JZ, Baleydier F, L'Heureux S, Wang H, Gordon WR, Thoryk E, Andrawes MB, Tiyanont K, Stegmaier K, Roti G, Ross KN, Franlin LL, Wang F, Chastain M, Bett AJ, Audoly LP, Aster JC, Blacklow SC, Huber HE (2010) Characterization of notch1 antibodies that inhibit signaling of both normal and mutated notch1 receptors. PLoS One 5:e9094

Beverly LJ, Felsher DW, Capobianco AJ (2005) Suppression of p53 by notch in lymphomagenesis: implications for initiation and regression. Cancer Res 65:7159–7168

Bhavsar PJ, Infante E, Khwaja A, Ridley AJ (2012) Analysis of Rho GTPase expression in T-ALL identifies RhoU as a target for notch involved in T-ALL cell migration. Oncogene.

Blaumueller CM, Qi H, Zagouras P, Artavanis-Tsakonas S (1997) Intracellular cleavage of Notch leads to a heterodimeric receptor on the plasma membrane. Cell 90:281–291

Bozkulak EC, Weinmaster G (2009) Selective use of ADAM10 and ADAM17 in activation of Notch1 signaling. Mol Cell Biol 29:5679–5695
Breit S, Stanulla M, Flohr T, Schrappe M, Ludwig WD, Tolle G, Happich M, Muckenthaler MU, Kulozik AE (2006) Activating NOTCH1 mutations predict favorable early treatment response and long-term outcome in childhood precursor T-cell lymphoblastic leukemia. Blood 108:1151–1157
Buonamici S, Trimarchi T, Ruocco MG, Reavie L, Cathelin S, Mar BG, Klinakis A, Lukyanov Y, Tseng JC, Sen F, Gehrie E, Li M, Newcomb E, Zavadil J, Meruelo D, Lipp M, Ibrahim S, Efstratiadis A, Zagzag D, Bromberg JS, Dustin ML, Aifantis I (2009) CCR7 signalling as an essential regulator of CNS infiltration in T-cell leukaemia. Nature 459:1000–1004
Cerchietti LC, Ghetu AF, Zhu X, Da Silva GF, Zhong S, Matthews M, Bunting KL, Polo JM, Fares C, Arrowsmith CH, Yang SN, Garcia M, Coop A, Mackerell AD Jr, Prive GG, Melnick A (2010) A small-molecule inhibitor of BCL6 kills DLBCL cells in vitro and in vivo. Cancer Cell 17:400–411
Chan SM, Weng AP, Tibshirani R, Aster JC, Utz PJ (2007) Notch signals positively regulate activity of the mTOR pathway in T-cell acute lymphoblastic leukemia. Blood 110:278–286
Chiang MY, Xu L, Shestova O, Histen G, L'Heureux S, Romany C, Childs ME, Gimotty PA, Aster JC, Pear WS (2008) Leukemia-associated NOTCH1 alleles are weak tumor initiators but accelerate K-ras-initiated leukemia. J Clin Invest 118:3181–3194
Ciofani M, Zuniga-Pflucker JC (2005) Notch promotes survival of pre-T cells at the beta-selection checkpoint by regulating cellular metabolism. Nat Immunol 6:881–888
Clappier E, Collette S, Grardel N, Girard S, Suarez L, Brunie G, Kaltenbach S, Yakouben K, Mazingue F, Robert A, Boutard P, Plantaz D, Rohrlich P, van Vlierberghe P, Preudhomme C, Otten J, Speleman F, Dastugue N, Suciu S, Benoit Y, Bertrand Y, Cave H (2010) NOTCH1 and FBXW7 mutations have a favorable impact on early response to treatment, but not on outcome, in children with T-cell acute lymphoblastic leukemia (T-ALL) treated on EORTC trials 58881 and 58951. Leukemia 24:2023–2031
Cullion K, Draheim KM, Hermance N, Tammam J, Sharma VM, Ware C, Nikov G, Krishnamoorthy V, Majumder PK, Kelliher MA (2009) Targeting the Notch1 and mTOR pathways in a mouse T-ALL model. Blood 113:6172–6181
D'Altri T, Gonzalez J, Aifantis I, Espinosa L, Bigas A (2011) Hes1 expression and CYLD repression are essential events downstream of notch1 in T-cell leukemia. Cell Cycle 10:1031–1036
De Strooper B, Annaert W, Cupers P, Saftig P, Craessaerts K, Mumm JS, Schroeter EH, Schrijvers V, Wolfe MS, Ray WJ, Goate A, Kopan R (1999) A presenilin-1-dependent gamma-secretase-like protease mediates release of notch intracellular domain. Nature 398:518–522
Deangelo D, Stone R, Silverman L, Stock W, Attar E, Fearen I, Dallob A, Matthews C, Stone J, Freedman S, Aster JC (2006) A phase I clinical trial of the notch inhibitor MK-0752 in patients with T-cell acute lymphoblastic leukemia/lymphoma (T-ALL) and other leukemias. ASCO Annual Meeting Proceedings Part I. J Clin Oncol 24:6585
Deftos ML, He YW, Ojala EW, Bevan MJ (1998) Correlating notch signaling with thymocyte maturation. Immunity 9:777–786
Demarest RM, Dahmane N, Capobianco AJ (2011) Notch is oncogenic dominant in T-cell acute lymphoblastic leukemia. Blood 117:2901–2909
Dohda T, Maljukova A, Liu L, Heyman M, Grander D, Brodin D, Sangfelt O, Lendahl U (2007) Notch signaling induces SKP2 expression and promotes reduction of p27Kip1 in T-cell acute lymphoblastic leukemia cell lines. Exp Cell Res 313:3141–3152
Dunn CD, Sulis ML, Ferrando AA, Greenwald I (2010) A conserved tetraspanin subfamily promotes notch signaling in *Caenorhabditis elegans* and in human cells. Proc Natl Acad Sci U S A 107:5907–5912
Eguchi-Ishimae M, Eguchi M, Kempski H, Greaves M (2008) NOTCH1 mutation can be an early, prenatal genetic event in T-ALL. Blood 111:376–378
Ellisen LW, Bird J, West DC, Soreng AL, Reynolds TC, Smith SD, Sklar J (1991) TAN-1, the human homolog of the *Drosophila* notch gene, is broken by chromosomal translocations in T lymphoblastic neoplasms. Cell 66:649–661
Espinosa L, Cathelin S, D'Altri T, Trimarchi T, Statnikov A, Guiu J, Rodilla V, Ingles-Esteve J, Nomdedeu J, Bellosillo B, Besses C, Abdel-Wahab O, Kucine N, Sun SC, Song G, Mullighan

CC, Levine RL, Rajewsky K, Aifantis I, Bigas A (2010) The notch/Hes1 pathway sustains NF-kappaB activation through CYLD repression in T cell leukemia. Cancer Cell 18:268–281
Evin G, Sernee MF, Masters CL (2006) Inhibition of gamma-secretase as a therapeutic intervention for Alzheimer's disease: prospects, limitations and strategies. CNS Drugs 20:351–372
Ferrando A (2010) NOTCH mutations as prognostic markers in T-ALL. Leukemia 24:2003–2004
Ferrando AA (2009) The role of NOTCH1 signaling in T-ALL. Hematology Am Soc Hematol Educ Program 353–361
Ferrando AA, Neuberg DS, Staunton J, Loh ML, Huard C, Raimondi SC, Behm FG, Pui CH, Downing JR, Gilliland DG, Lander ES, Golub TR, Look AT (2002) Gene expression signatures define novel oncogenic pathways in T cell acute lymphoblastic leukemia. Cancer Cell 1:75–87
Fryer CJ, White JB, Jones KA (2004) Mastermind recruits CycC:CDK8 to phosphorylate the notch ICD and coordinate activation with turnover. Mol Cell 16:509–520
Girard L, Hanna Z, Beaulieu N, Hoemann CD, Simard C, Kozak CA, Jolicoeur P (1996) Frequent provirus insertional mutagenesis of notch1 in thymomas of MMTVD/myc transgenic mice suggests a collaboration of c-myc and notch1 for oncogenesis. Genes Dev 10:1930–1944
Gomez-del Arco P, Kashiwagi M, Jackson AF, Naito T, Zhang J, Liu F, Kee B, Vooijs M, Radtke F, Redondo JM, Georgopoulos K (2010) Alternative promoter usage at the notch1 locus supports ligand-independent signaling in T cell development and leukemogenesis. Immunity 33:685–698
Gonzalez-Garcia S, Garcia-Peydro M, Martin-Gayo E, Ballestar E, Esteller M, Bornstein R, de la Pompa JL, Ferrando AA, Toribio ML (2009) CSL-MAML-dependent notch1 signaling controls T lineage-specific IL-7R{alpha} gene expression in early human thymopoiesis and leukemia. J Exp Med 206:779–791
Gordon WR, Roy M, Vardar-Ulu D, Garfinkel M, Mansour MR, Aster JC, Blacklow SC (2009) Structure of the notch1-negative regulatory region: implications for normal activation and pathogenic signaling in T-ALL. Blood 113:4381–4390
Hsieh JJ, Henkel T, Salmon P, Robey E, Peterson MG, Hayward SD (1996) Truncated mammalian notch1 activates CBF1/RBPJk-repressed genes by a mechanism resembling that of epstein-barr virus EBNA2. Mol Cell Biol 16:952–959
Jeannet R, Mastio J, Macias-Garcia A, Oravecz A, Ashworth T, Geimer Le Lay AS, Jost B, Le Gras S, Ghysdael J, Gridley T, Honjo T, Radtke F, Aster JC, Chan S, Kastner P (2010) Oncogenic activation of the notch1 gene by deletion of its promoter in Ikaros-deficient T-ALL. Blood 116:5443–5454
Jensen J, Pedersen EE, Galante P, Hald J, Heller RS, Ishibashi M, Kageyama R, Guillemot F, Serup P, Madsen OD (2000) Control of endodermal endocrine development by Hes-1. Nat Genet 24:36–44
Joshi I, Minter LM, Telfer J, Demarest RM, Capobianco AJ, Aster JC, Sicinski P, Fauq A, Golde TE, Osborne BA (2009) Notch signaling mediates G1/S cell-cycle progression in T cells via cyclin D3 and its dependent kinases. Blood 113:1689–1698
Katz JP, Perreault N, Goldstein BG, Lee CS, Labosky PA, Yang VW, Kaestner KH (2002) The zinc-finger transcription factor Klf4 is required for terminal differentiation of goblet cells in the colon. Development 129:2619–2628
Kim TH, Shivdasani RA (2011) Genetic evidence that intestinal notch functions vary regionally and operate through a common mechanism of math1 repression. J Biol Chem 286:11427–11433
Klinakis A, Lobry C, Abdel-Wahab O, Oh P, Haeno H, Buonamici S, van De Walle I, Cathelin S, Trimarchi T, Araldi E, Liu C, Ibrahim S, Beran M, Zavadil J, Efstratiadis A, Taghon T, Michor F, Levine RL, Aifantis I (2011) A novel tumour-suppressor function for the notch pathway in myeloid leukaemia. Nature 473:230–233
Kox C, Zimmermann M, Stanulla M, Leible S, Schrappe M, Ludwig WD, Koehler R, Tolle G, Bandapalli OR, Breit S, Muckenthaler MU, Kulozik AE (2010) The favorable effect of activating NOTCH1 receptor mutations on long-term outcome in T-ALL patients treated on the ALL-BFM 2000 protocol can be separated from FBXW7 loss of function. Leukemia 24:2005–2013
Liu H, Chi AW, Arnett KL, Chiang MY, Xu L, Shestova O, Wang H, Li YM, Bhandoola A, Aster JC, Blacklow SC, Pear WS (2010) Notch dimerization is required for leukemogenesis and T-cell development. Genes Dev 24:2395–2407

Logeat F, Bessia C, Brou C, LeBail O, Jarriault S, Seidah NG, Israel A (1998) The notch1 receptor is cleaved constitutively by a furin-like convertase. Proc Natl Acad Sci U S A 95:8108–8112

Malecki MJ, Sanchez-Irizarry C, Mitchell JL, Histen G, Xu ML, Aster JC, Blacklow SC (2006) Leukemia-associated mutations within the NOTCH1 heterodimerization domain fall into at least two distinct mechanistic classes. Mol Cell Biol 26:4642–4651

Mansour MR, Duke V, Foroni L, Patel B, Allen CG, Ancliff PJ, Gale RE, Linch DC (2007) Notch-1 mutations are secondary events in some patients with T-cell acute lymphoblastic leukemia. Clin Cancer Res 13:6964–6969

Mansour MR, Sulis ML, Duke V, Foroni L, Jenkinson S, Koo K, Allen CG, Gale RE, Buck G, Richards S, Paietta E, Rowe JM, Tallman MS, Goldstone AH, Ferrando AA, Linch DC (2009) Prognostic implications of NOTCH1 and FBXW7 mutations in adults with T-cell acute lymphoblastic leukemia treated on the MRC UKALLXII/ECOG E2993 protocol. J Clin Oncol 27:4352–4356

Margolin AA, Palomero T, Sumazin P, Califano A, Ferrando AA, Stolovitzky G (2009) ChIP-on-chip significance analysis reveals large-scale binding and regulation by human transcription factor oncogenes. Proc Natl Acad Sci U S A 106:244–249

Medyouf H, Gao X, Armstrong F, Gusscott S, Liu Q, Gedman AL, Matherly LH, Schultz KR, Pflumio F, You MJ, Weng AP (2010) Acute T-cell leukemias remain dependent on notch signaling despite PTEN and INK4A/ARF loss. Blood 115:1175–1184

Medyouf H, Gusscott S, Wang H, Tseng JC, Wai C, Nemirovsky O, Trumpp A, Pflumio F, Carboni J, Gottardis M, Pollak M, Kung AL, Aster JC, Holzenberger M, Weng AP (2011) High-level IGF1R expression is required for leukemia-initiating cell activity in T-ALL and is supported by notch signaling. J Exp Med 208:1809–1822

Minella AC, Clurman BE (2005) Mechanisms of tumor suppression by the SCF(Fbw7). Cell Cycle 4:1356–1359

Mirandola L, Chiriva-Internati M, Montagna D, Locatelli F, Zecca M, Ranzani M, Basile A, Locati M, Cobos E, Kast WM, Asselta R, Paraboschi EM, Comi P, Chiaramonte R (2012) Notch1 regulates chemotaxis and proliferation by controlling the CC-chemokine receptors 5 and 9 in T cell acute lymphoblastic leukaemia. J Pathol 226:713–722

Mo JS, Ann EJ, Yoon JH, Jung J, Choi YH, Kim HY, Ahn JS, Kim SM, Kim MY, Hong JA, Seo MS, Lang F, Choi EJ, Park HS (2011) Serum- and glucocorticoid-inducible kinase 1 (SGK1) controls notch1 signaling by downregulation of protein stability through Fbw7 ubiquitin ligase. J Cell Sci 124:100–112

Moellering RE, Cornejo M, Davis TN, Del Bianco C, Aster JC, Blacklow SC, Kung AL, Gilliland DG, Verdine GL, Bradner JE (2009) Direct inhibition of the NOTCH transcription factor complex. Nature 462:182–188

Mumm JS, Schroeter EH, Saxena MT, Griesemer A, Tian X, Pan DJ, Ray WJ, Kopan R (2000) A ligand-induced extracellular cleavage regulates gamma-secretase-like proteolytic activation of notch1. Mol Cell 5:197–206

O'Neil J, Grim J, Strack P, Rao S, Tibbitts D, Winter C, Hardwick J, Welcker M, Meijerink JP, Pieters R, Draetta G, Sears R, Clurman BE, Look AT (2007) FBW7 mutations in leukemic cells mediate NOTCH pathway activation and resistance to gamma-secretase inhibitors. J Exp Med 204:1813–1824

Palomero T, Barnes KC, Real PJ, Bender JL, Sulis ML, Murty VV, Colovai AI, Balbin M, Ferrando AA (2006a) CUTLL1, a novel human T-cell lymphoma cell line with t(7;9) rearrangement, aberrant NOTCH1 activation and high sensitivity to gamma-secretase inhibitors. Leukemia 20:1279–1287

Palomero T, Lim WK, Odom DT, Sulis ML, Real PJ, Margolin A, Barnes KC, O'Neil J, Neuberg D, Weng AP, Aster JC, Sigaux F, Soulier J, Look AT, Young RA, Califano A, Ferrando AA (2006b) NOTCH1 directly regulates c-MYC and activates a feed-forward-loop transcriptional network promoting leukemic cell growth. Proc Natl Acad Sci U S A 103:18261–18266

Palomero T, Sulis ML, Cortina M, Real PJ, Barnes K, Ciofani M, Caparros E, Buteau J, Brown K, Perkins SL, Bhagat G, Agarwal AM, Basso G, Castillo M, Nagase S, Cordon-Cardo C,

Parsons R, Zuniga-Pflucker JC, Dominguez M, Ferrando AA (2007) Mutational loss of PTEN induces resistance to NOTCH1 inhibition in T-cell leukemia. Nat Med 13:1203–1210

Park MJ, Taki T, Oda M, Watanabe T, Yumura-Yagi K, Kobayashi R, Suzuki N, Hara J, Horibe K, Hayashi Y (2009) FBXW7 and NOTCH1 mutations in childhood T cell acute lymphoblastic leukaemia and T cell non-Hodgkin lymphoma. Br J Haematol 145:198–206

Pear WS, Aster JC, Scott ML, Hasserjian RP, Soffer B, Sklar J, Baltimore D (1996) Exclusive development of T cell neoplasms in mice transplanted with bone marrow expressing activated Notch alleles. J Exp Med 183:2283–2291

Porcu M, Kleppe M, Gianfelici V, Geerdens E, De Keersmaecker K, Tartaglia M, Foa R, Soulier J, Cauwelier B, Uyttebroeck A, Macintyre E, Vandenberghe P, Asnafi V, Cools J (2012) Mutation of the receptor tyrosine phosphatase PTPRC (CD45) in T-cell acute lymphoblastic leukemia. Blood

Puente XS, Pinyol M, Quesada V, Conde L, Ordonez GR, Villamor N, Escaramis G, Jares P, Bea S, Gonzalez-Diaz M, Bassaganyas L, Baumann T, Juan M, Lopez-Guerra M, Colomer D, Tubio JM, Lopez C, Navarro A, Tornador C, Aymerich M, Rozman M, Hernandez JM, Puente DA, Freije JM, Velasco G, Gutierrez-Fernandez A, Costa D, Carrio A, Guijarro S, Enjuanes A, Hernandez L, Yague J, Nicolas P, Romeo-Casabona CM, Himmelbauer H, Castillo E, Dohm JC, de Sanjose S, Piris MA, de Alava E, San Miguel J, Royo R, Gelpi JL, Torrents D, Orozco M, Pisano DG, Valencia A, Guigo R, Bayes M, Heath S, Gut M, Klatt P, Marshall J, Raine K, Stebbings LA, Futreal PA, Stratton MR, Campbell PJ, Gut I, Lopez-Guillermo A, Estivill X, Montserrat E, Lopez-Otin C, Campo E (2011) Whole-genome sequencing identifies recurrent mutations in chronic lymphocytic leukaemia. Nature 475:101–105

Pui JC, Allman D, Xu L, DeRocco S, Karnell FG, Bakkour S, Lee JY, Kadesch T, Hardy RR, Aster JC, Pear WS (1999) Notch1 expression in early lymphopoiesis influences B versus T lineage determination. Immunity 11:299–308

Radtke F, Wilson A, Stark G, Bauer M, van Meerwijk J, MacDonald HR, Aguet M (1999) Deficient T cell fate specification in mice with an induced inactivation of notch1. Immunity 10:547–558

Rand MD, Grimm LM, Artavanis-Tsakonas S, Patriub V, Blacklow SC, Sklar J, Aster JC (2000) Calcium depletion dissociates and activates heterodimeric notch receptors. Mol Cell Biol 20:1825–1835

Rao SS, O'Neil J, Liberator CD, Hardwick JS, Dai X, Zhang T, Tyminski E, Yuan J, Kohl NE, Richon VM, Van der Ploeg LH, Carroll PM, Draetta GF, Look AT, Strack PR, Winter CG (2009) Inhibition of NOTCH signaling by gamma secretase inhibitor engages the RB pathway and elicits cell cycle exit in T-cell acute lymphoblastic leukemia cells. Cancer Res 69:3060–3068

Real PJ, Tosello V, Palomero T, Castillo M, Hernando E, de Stanchina E, Sulis ML, Barnes K, Sawai C, Homminga I, Meijerink J, Aifantis I, Basso G, Cordon-Cardo C, Ai W, Ferrando A (2009) Gamma-secretase inhibitors reverse glucocorticoid resistance in T cell acute lymphoblastic leukemia. Nat Med 15:50–58

Rebay I, Fleming RJ, Fehon RG, Cherbas L, Cherbas P, Artavanis-Tsakonas S (1991) Specific EGF repeats of notch mediate interactions with delta and serrate: implications for notch as a multifunctional receptor. Cell 67:687–699

Rechsteiner M (1988) Regulation of enzyme levels by proteolysis: the role of pest regions. Adv Enzyme Regul 27:135–151

Riccio O, van Gijn ME, Bezdek AC, Pellegrinet L, van Es JH, Zimber-Strobl U, Strobl LJ, Honjo T, Clevers H, Radtke F (2008) Loss of intestinal crypt progenitor cells owing to inactivation of both notch1 and notch2 is accompanied by derepression of CDK inhibitors p27(Kip1) and p57(Kip2). EMBO Rep 9:377–383

Sade H, Krishna S, Sarin A (2004) The anti-apoptotic effect of notch-1 requires p56lck-dependent, Akt/PKB-mediated signaling in T cells. J Biol Chem 279:2937–2944

Sanchez-Irizarry C, Carpenter AC, Weng AP, Pear WS, Aster JC, Blacklow SC (2004) Notch subunit heterodimerization and prevention of ligand-independent proteolytic activation depend, respectively, on a novel domain and the LNR repeats. Mol Cell Biol 24:9265–9273

Schroeter EH, Kisslinger JA, Kopan R (1998) Notch-1 signalling requires ligand-induced proteolytic release of intracellular domain. Nature 393:382–386

Selkoe D, Kopan R (2003) Notch and Presenilin: regulated intramembrane proteolysis links development and degeneration. Annu Rev Neurosci 26:565–597

Sharma VM, Calvo JA, Draheim KM, Cunningham LA, Hermance N, Beverly L, Krishnamoorthy V, Bhasin M, Capobianco AJ, Kelliher MA (2006) Notch1 contributes to mouse T-cell leukemia by directly inducing the expression of c-myc. Mol Cell Biol 26:8022–8031

Shin HM, Minter LM, Cho OH, Gottipati S, Fauq AH, Golde TE, Sonenshein GE, Osborne BA (2006) Notch1 augments NF-kappaB activity by facilitating its nuclear retention. Embo J 25:129–138

Shochat C, Tal N, Bandapalli OR, Palmi C, Ganmore I, te Kronnie G, Cario G, Cazzaniga G, Kulozik AE, Stanulla M, Schrappe M, Biondi A, Basso G, Bercovich D, Muckenthaler MU, Izraeli S (2011) Gain-of-function mutations in interleukin-7 receptor-alpha (IL7R) in childhood acute lymphoblastic leukemias. J Exp Med 208:901–908

Song LL, Peng Y, Yun J, Rizzo P, Chaturvedi V, Weijzen S, Kast WM, Stone PJ, Santos L, Loredo A, Lendahl U, Sonenshein G, Osborne B, Qin JZ, Pannuti A, Nickoloff BJ, Miele L (2008) Notch-1 associates with IKKalpha and regulates IKK activity in cervical cancer cells. Oncogene 27:5833–5844

Struhl G, Greenwald I (1999) Presenilin is required for activity and nuclear access of Notch in *Drosophila*. Nature 398:522–525

Sulis ML, Saftig P, Ferrando AA (2011) Redundancy and specificity of the metalloprotease system mediating oncogenic NOTCH1 activation in T-ALL. Leukemia 25:1564–1569

Sulis ML, Williams O, Palomero T, Tosello V, Pallikuppam S, Real PJ, Barnes K, Zuurbier L, Meijerink JP, Ferrando AA (2008) NOTCH1 extracellular juxtamembrane expansion mutations in T-ALL. Blood 112:733–740

Tammam J, Ware C, Efferson C, O'Neil J, Rao S, Qu X, Gorenstein J, Angagaw M, Kim H, Kenific C, Kunii K, Leach KJ, Nikov G, Zhao J, Dai X, Hardwick J, Scott M, Winter C, Bristow L, Elbi C, Reilly JF, Look T, Draetta G, Van der Ploeg L, Kohl NE, Strack PR, Majumder PK (2009) Down-regulation of the Notch pathway mediated by a gamma-secretase inhibitor induces anti-tumour effects in mouse models of T-cell leukaemia. Br J Pharmacol 158:1183–1195

Tatarek J, Cullion K, Ashworth T, Gerstein R, Aster JC, Kelliher MA (2011) Notch1 inhibition targets the leukemia-initiating cells in a Tal1/Lmo2 mouse model of T-ALL. Blood 118:1579–1590

Thompson BJ, Buonamici S, Sulis ML, Palomero T, Vilimas T, Basso G, Ferrando A, Aifantis I (2007) The SCFFBW7 ubiquitin ligase complex as a tumor suppressor in T cell leukemia. J Exp Med 204:1825–1835

Tsuji H, Ishii-Ohba H, Ukai H, Katsube T, Ogiu T (2003) Radiation-induced deletions in the 5′ end region of notch1 lead to the formation of truncated proteins and are involved in the development of mouse thymic lymphomas. Carcinogenesis 24:1257–1268

van Es JH, van Gijn ME, Riccio O, van den Born M, Vooijs M, Begthel H, Cozijnsen M, Robine S, Winton DJ, Radtke F, Clevers H (2005) Notch/gamma-secretase inhibition turns proliferative cells in intestinal crypts and adenomas into goblet cells. Nature 435:959–963

van Grotel M, Meijerink JP, Beverloo HB, Langerak AW, Buys-Gladdines JG, Schneider P, Poulsen TS, den Boer ML, Horstmann M, Kamps WA, Veerman AJ, van Wering ER, van Noesel MM, Pieters R (2006) The outcome of molecular-cytogenetic subgroups in pediatric T-cell acute lymphoblastic leukemia: a retrospective study of patients treated according to DCOG or COALL protocols. Haematologica 91:1212–1221

van Tetering G, van Diest P, Verlaan I, van der Wall E, Kopan R, Vooijs M (2009) Metalloprotease ADAM10 is required for notch1 site 2 cleavage. J Biol Chem 284:31018–31027

von Boehmer H, Aifantis I, Azogui O, Feinberg J, Saint-Ruf C, Zober C, Garcia C, Buer J (1998) Crucial function of the pre-T-cell receptor (TCR) in TCR beta selection, TCR beta allelic exclusion and alpha beta versus gamma delta lineage commitment. Immunol Rev 165:111–119

Wang NJ, Sanborn Z, Arnett KL, Bayston LJ, Liao W, Proby CM, Leigh IM, Collisson EA, Gordon PB, Jakkula L, Pennypacker S, Zou Y, Sharma M, North JP, Vemula SS, Mauro TM, Neuhaus IM, Leboit PE, Hur JS, Park K, Huh N, Kwok PY, Arron ST, Massion PP, Bale AE, Haussler D, Cleaver JE, Gray JW, Spellman PT, South AP, Aster JC, Blacklow SC, Cho RJ

(2011) Loss-of-function mutations in Notch receptors in cutaneous and lung squamous cell carcinoma. Proc Natl Acad Sci U S A 108:17761–17766

Wei P, Walls M, Qiu M, Ding R, Denlinger RH, Wong A, Tsaparikos K, Jani JP, Hosea N, Sands M, Randolph S, Smeal T (2010) Evaluation of selective gamma-secretase inhibitor PF-03084014 for its antitumor efficacy and gastrointestinal safety to guide optimal clinical trial design. Mol Cancer Ther 9:1618–1628

Wendorff AA, Koch U, Wunderlich FT, Wirth S, Dubey C, Bruning JC, MacDonald HR, Radtke F (2010) Hes1 is a critical but context-dependent mediator of canonical notch signaling in lymphocyte development and transformation. Immunity 33:671–684

Weng AP,Ferrando AA,Lee W, Morris JPt, Silverman LB, Sanchez-Irizarry C, Blacklow SC, Look AT, Aster JC (2004) Activating mutations of NOTCH1 in human T cell acute lymphoblastic leukemia. Science 306:269–271

Weng AP, Millholland JM, Yashiro-Ohtani Y, Arcangeli ML, Lau A, Wai C, Del Bianco C, Rodriguez CG, Sai H, Tobias J, Li Y, Wolfe MS, Shachaf C, Felsher D, Blacklow SC, Pear WS, Aster JC (2006) c-Myc is an important direct target of notch1 in T-cell acute lymphoblastic leukemia/lymphoma. Genes Dev 20:2096–2109

Weng AP, Nam Y, Wolfe MS, Pear WS, Griffin JD, Blacklow SC, Aster JC (2003) Growth suppression of pre-T acute lymphoblastic leukemia cells by inhibition of notch signaling. Mol Cell Biol 23:655–664

Westhoff B, Colaluca IN, D'Ario G, Donzelli M, Tosoni D, Volorio S, Pelosi G, Spaggiari L, Mazzarol G, Viale G, Pece S, Di Fiore PP (2009) Alterations of the notch pathway in lung cancer. Proc Natl Acad Sci U S A 106:22293–22298

Wu L, Aster JC, Blacklow SC, Lake R, Artavanis-Tsakonas S, Griffin JD (2000) MAML1, a human homologue of *Drosophila* mastermind, is a transcriptional co-activator for NOTCH receptors. Nat Genet 26:484–489

Wu Y, Cain-Hom C, Choy L, Hagenbeek TJ, de Leon GP, Chen Y, Finkle D, Venook R, Wu X, Ridgway J, Schahin-Reed D, Dow GJ, Shelton A, Stawicki S, Watts RJ, Zhang J, Choy R, Howard P, Kadyk L, Yan M, Zha J, Callahan CA, Hymowitz SG, Siebel CW (2010) Therapeutic antibody targeting of individual notch receptors. Nature 464:1052–1057

Yang Q, Bermingham NA, Finegold MJ, Zoghbi HY (2001) Requirement of Math1 for secretory cell lineage commitment in the mouse intestine. Science 294:2155–2158

Zenatti PP, Ribeiro D, Li W, Zuurbier L, Silva MC, Paganin M, Tritapoe J, Hixon JA, Silveira AB, Cardoso BA, Sarmento LM, Correia N, Toribio ML, Kobarg J, Horstmann M, Pieters R, Brandalise SR, Ferrando AA, Meijerink JP, Durum SK, Yunes JA, Barata JT (2011) Oncogenic IL7R gain-of-function mutations in childhood T-cell acute lymphoblastic leukemia. Nat Genet 43:932–939

Zhang J, Ding L, Holmfeldt L, Wu G, Heatley SL, Payne-Turner D, Easton J, Chen X, Wang J, Rusch M, Lu C, Chen SC, Wei L, Collins-Underwood JR, Ma J, Roberts KG, Pounds SB, Ulyanov A, Becksfort J, Gupta P, Huether R, Kriwacki RW, Parker M, McGoldrick DJ, Zhao D, Alford D, Espy S, Bobba KC, Song G, Pei D, Cheng C, Roberts S, Barbato MI, Campana D, Coustan-Smith E, Shurtleff SA, Raimondi SC, Kleppe M, Cools J, Shimano KA, Hermiston ML, Doulatov S, Eppert K, Laurenti E, Notta F, Dick JE, Basso G, Hunger SP, Loh ML, Devidas M, Wood B, Winter S, Dunsmore KP, Fulton RS, Fulton LL, Hong X, Harris CC, Dooling DJ, Ochoa K, Johnson KJ, Obenauer JC, Evans WE, Pui CH, Naeve CW, Ley TJ, Mardis ER, Wilson RK, Downing JR, Mullighan CG (2012) The genetic basis of early T-cell precursor acute lymphoblastic leukaemia. Nature 481:157–163

Zuurbier L, Homminga I, Calvert V, te Winkel ML, Buijs-Gladdines JG, Kooi C, Smits WK, Sonneveld E, Veerman AJ, Kamps WA, Horstmann M, Petricoin EF 3rd, Pieters R, Meijerink JP (2010) NOTCH1 and/or FBXW7 mutations predict for initial good prednisone response but not for improved outcome in pediatric T-cell acute lymphoblastic leukemia patients treated on DCOG or COALL protocols. Leukemia 24:2014–2022

Index

Current Topics in Microbiology and Immunology (2012) 360: 183–184
DOI: 10.1007/978-3-642-24294-6